Selected Topics in RF, Analog and Mixed Signal Circuits and Systems

EDITORS

Kiran Gunnam
Western Digital Corporation, USA

Mohammad (Vahid) VahidFar
Apple, USA

Tutorials in Circuits and Systems

For a list of other books in this series, visit www.riverpublishers.com

Series Editor
Franco Maloberti
President IEEE CAS Society
University of Pavia, Italy

LONDON AND NEW YORK

Published 2017 by River Publishers

River Publishers

Alsbjergvej 10, 9260 Gistrup, Denmark

www.riverpublishers.com

Distributed exclusively by Routledge

4 Park Square, Milton Park, Abingdon, Oxon OX14 4RN

605 Third Avenue, New York, NY 10158

First published in paperback 2024

Selected Topics in RF, Analog and Mixed Signal Circuits and Systems / by Kiran Gunnam, Mohammad (Vahid) VahidFar, Franco Maloberti.

Routledge is an imprint of the Taylor & Francis Group, an informa business

Publisher's Note
The publisher has gone to great lengths to ensure the quality of this reprint but points out that some imperfections in the original copies may be apparent.

While every effort is made to provide dependable information, the publisher, authors, and editors cannot be held responsible for any errors or omissions.

ISBN: 978-87-93519-18-3 (hbk)
ISBN: 978-87-7004-443-1 (pbk)
ISBN: 978-1-003-33944-1 (ebk)

DOI: 10.1201/9781003339441

Table of contents

Introduction

Advancements in solid-state circuits and systems continue to drive the ongoing innovations and smart applications around us in daily life basis. The communication revolution with high speed Internet and high-speed wireless connectivity is an explicit example of year-to-year speed increase due to the advancement in new standards and technologies. The new applications and even higher speed are coming soon which are expected to leverage on the broadening frequency-spectrum access from radio frequencies to millimeter-wave frequencies. Thus, more available bandwidth leads to the proportionate increase of Shannon channel capacity. Shorter wavelengths at mm-Waves enables compact realization of multi-antenna beam-forming transceivers with dense frequency reuse.

The data traffic and data-center workloads are facing outstripping capacity as every one needs to access many pages online and many social networks are connecting billions of people. High rate wire-line and optical data connectivity and fast CPU/GPU processing are essential part of this traffic handling beside all the cloud computing and storage. The exploding bandwidth requirements in this interconnect space clearly demand a higher interface speed like Terabit Ethernet to improve the efficiency at which we acquire, network, store, and process information

CMOS technology is at the heart of many recent developments in the design of integrated circuits. Moore's Law has served as the guiding principle for the semiconductor industry for several years. This trend is still moving forward as the state-of-the-art sub nm scaled CMOS technologies, for applications ranging from high-performance computing down to ultra-low-power mobile applications are developing. Circuit and system designers all around the globe are leveraging the immense device density and processing power of modern technology toward new applications for more smart and interconnected world.

This work aims to provide state of the art overview on the matter of high performance integrated circuits and systems design. In this book several advanced topics in the area of RF, Analog and Mixed Signal Circuits and Systems have been selected to be explored and to dig into their fundamentals and to present the state of the art developments. The book will cover ADC conversion and equalization for high-speed links, clock and data recovery for Gb/s link, signal generation for terra hertz application, and oscillator phase noise and analog/digital PLLs.

The first chapter is about the Oscillator Phase Noise. Over the last 20 years, the analysis of phase noise in LC oscillators has been one of the most discussed topics in the area of RF IC design. Phase noise is difficult to analyze and represents one of the main bottlenecks in the design of a transceiver.

Linear time-invariant (LTI) phase noise theories provide important qualitative design insights but are limited in their quantitative predictive power. Part of the difficulty is that device noise undergoes multiple frequency translations to become oscillator phase noise. A quantitative understanding of this process requires abandoning the principle of time invariance assumed in older theories of phase noise. Fortunately, the noise-to-phase transfer function of oscillators is still linear, despite the existence of the nonlinearities necessary for amplitude stabilization. In addition to providing a quantitative reconciliation between theory and measurement, the time-varying phase-noise model presented in this chapter identifies the importance of symmetry in suppressing the up conversion of noise into close-in phase noise, and provides an explicit appreciation of cyclostationary effects and AM–PM conversion. These insights allow a reinterpretation of why the Colpitts oscillator exhibits good performance, and suggest new oscillator topologies. Tuned LC and ring oscillator circuit examples are presented to reinforce the theoretical considerations developed. Simulation issues and the accommodation of amplitude noise are explained.

The second chapter is assigned to Clock and Data Recovery in High-Speed Wireline Communication. The chapter reviews the basics of timing and clock and data recovery circuits in high speed links, including system level considerations, limitations of the linear model, pull-in process and false lock, and a case for and key features of time domain event-driven system model to capture the effect of the circuit parameters on system performance. The circuits for phase detection and modeling of samplers for fast and accurate system level modeling is studied in details.

The third chapter is an overview of PLL design techniques to an IC designer. Along with basic feedback loop theory and common circuit implementations, the chapter provides details on Integer/Frac-N, CMOS delay cell ring and LC VCO based PLL designs. Some details on current trends in Dual Loop control and Digital PLL are also provided. Details on calibration and on-chip performance monitoring as well as Jitter analysis and measurement techniques are described.

The fourth chapter deals with Terahertz and mm-Wave Signal Generation, Synthesis and Amplification with an overview on the fundamental limits.

There is a growing interest in terahertz and mm-wave systems for compact, low cost and energy efficient imaging, spectroscopy and high data rate communication. Unfortunately, today's solid-state technologies including silicon and compound semiconductors have much lower performance at mm-wave frequencies compared to traditional RF bands. In order to overcome this limitation, we have introduced systematic methodologies for designing circuits and components, such as signal sources and amplifiers operating close to and beyond the conventional limits of the devices. These circuit blocks can effectively generate and combine signals from multiple devices to achieve performances orders of magnitude better than the state of the art. As an example, the implementation of a 482 GHz oscillator with an output power of 160 uW (-7.9 dBm) in 65 nm CMOS, a 300 GHz frequency synthesizer with 7.9% locking range in 90 nm SiGe, and a 260 GHz amplifier with a gain of 9.2 dB and saturated output power of -3.9 dBm in 65 nm CMOS is shown.

The fifth chapter covers Equalization and ADC conversion for high-speed links. As modern electrical and optical communication systems transition toward advanced modulation schemes, there exists a pressing need for use of power efficient A/D for serial interfaces operating above 28Gbps. Within this context, this chapter will cover architecture and circuit level design techniques for the front-end circuitry of ADC-based high-speed link receivers. The first part of the chapter will focus on system-level considerations, and will establish a connection between bit error rates and ADC resolution requirements. The second part of the chapter will look into the opportunity of relaxing the ADC requirements using analog pre-equalization. Finally, we will discuss the basics of implementing the required A/D converters, with an emphasis on time interleaved architectures.

CHAPTER 1

Linearity, Time-Variation, Phase Modulation and Oscillator Phase Noise

Thomas H. Lee

Stanford University, USA

Linear time-invariant (LTI) phase noise theories provide important qualitative design insights but are limited in their quantitative predictive power. Part of the difficulty is that device noise undergoes multiple frequency translations to become oscillator phase noise. A quantitative understanding of this process requires abandoning the principle of time invariance assumed in most older theories of phase noise. Fortunately, the noise-to-phase transfer function of oscillators is still linear, despite the existence of the nonlinearities necessary for amplitude stabilization. In addition to providing a quantitative reconciliation between theory and measurement, the time-varying phase-noise model presented in this tutorial identifies the importance of symmetry in suppressing the upconversion of 1 noise into close-in phase noise, and provides an explicit appreciation of cyclostationary effects and AM–PM conversion. These insights allow a reinterpretation of why the Colpitts oscillator exhibits good performance, and suggest new oscillator topologies. Tuned LC and ring oscillator circuit examples are presented to reinforce the theoretical considerations developed. Simulation issues and the accommodation of amplitude noise are considered in appendixes.

1 **Preliminaries (to refresh dormant neurons)**

- A system is *linear* if superposition holds.
  - Scaling of a single input is included.
  - Impulse response yields sufficient information to deduce the response to an arbitrary input.
  - All real systems can be driven into nonlinearity.
    - Linearity only holds over limited range.
- A system is *time-invariant* if time-shifting an input *only* time-shifts the output.
- If a system is LTI, then excitation at *f* produces steady-state response only at *f*.

Before we start the analysis, it is good idea to have a review on some of the signal and system fundamentals. Let's start with linear system in which a linear system hold the superposition for the multiple inputs applied to the system; the output of linear system is scaled with the input amplitude. More importantly, Impulse response yields sufficient information to deduce the response to an arbitrary input. Although all the real systems can be driven to the non-linear region, however linearity can be applied for the limited range and model the system as a linear system for that particular region. A system is considered time-invariant If any time shift at the input signal cause only the same time shift at the output signal. If a system is LTI (Linear Time Invariant), then excitation at frequency of f produces steady-state response only at the same frequency. In the other words a LTI system can't generate frequencies other than those frequencies are applied at the input.

2 **Preliminaries**

- If a system is LTV, excitation at *f* can produce steady-state response at other than *f*.
  - Superposition holds, so impulse response *still* tells us about response to any other input.

- If a system is nonlinear, excitation at *f* can also produce response at other than *f*.
  - Superposition doesn't hold; impulse response cannot be used to infer response to arbitrary excitations.

On the other hand for a LTV (Linear time variant) system, excitation at frequency of f can produce steady-state response at frequencies other than f. Like LTI system, superposition holds, so impulse response still tells us about response to any other input. Considering a non-linear system, the excitation at a frequency f can also produce response at frequencies other than f. As an example a non-linear system can produce harmonics of the frequency at 2f, 3f and etc. However the impulse response of a non-linear system cannot be used to predict the output for any arbitrary input; In the other words unlike LTI system, superposition doesn't hold in a non-linear system.

Oscillator have inputs, not just outputs 3

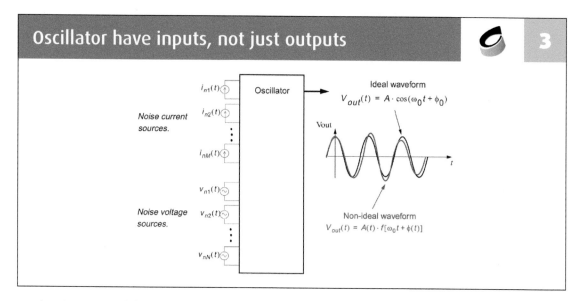

After this review of the linear and non-linear system, let's start looking at an oscillator; the oscillator usually assumed to have no inputs; However looking more carefully it is seen that the operation of an oscillator, the source noise are important to consider, or even the oscillator to starts; As shown in the figure; there are several current and voltage sources in every oscillator; Like for any LC oscillator the loss at the inductor or capacitor can be modeled by resistor and the noise of that resistor by either a current or voltage source. Looking at the oscillator output, it is ideally a waveform by a constant amplitude or phase; however the real oscillator could have both time variant amplitude and phase as it is shown. The time variant phase is the focus of our discussion and is referred as phase noise.

Phase noise effects in RF systems 4

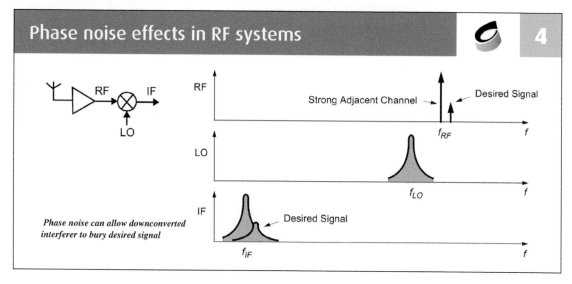

Now the question could be, why we care so much for the oscillator phase noise; To understand that lets look at a an RF receiver which is a common block in all our cell phones. As it is shown, the receiver job is to translate the high frequency input signal to a high quality low frequency signal like audio in a phone conversation. However the spectrum is not clean and there might be interferes close to the desired signal by some order of magnitude higher power; Usually the receiver should filter out the undesired signal however if the local oscillator (LO) is not clean and has poor phase noise, it can down converted the interfere to the desired signal band and bury the desired signal;

5 Units of phase noise

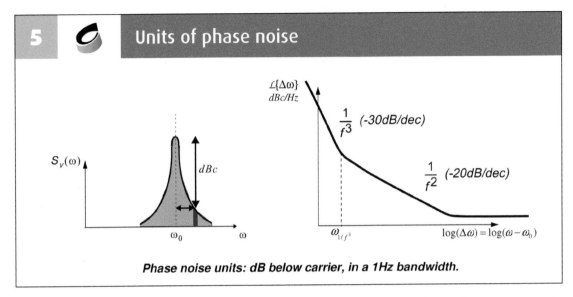

Phase noise units: dB below carrier, in a 1Hz bandwidth.

L ets look more closely at the spectrum of the LO with phase noise. Let's assume the oscillator frequency is f0, then the phase noise power at each offset frequency from f0 is defined by the ratio of the power per hertz at that offset frequency to the total signal power. As it is shown, usually the phase noise is worse at lower offset frequencies; The graph shows the typical phase noise spectrum for an LC oscillator in which the phase noise power reduces by -20dB/decade before get limited by thermal noise.

6 Substrate and supply noise matter, too

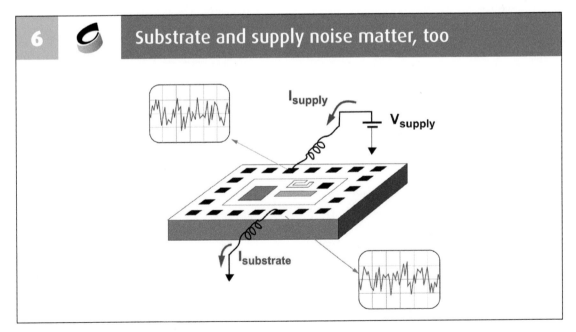

I n any system, there are several noise sources that can affect the oscillator performance including the noise from supply and ground network. Any current variation can introduce ripples due to the inductive part of the supply routing.

Phase noise: General considerations

  7

- Idealize oscillator as RLC + negative resistance:

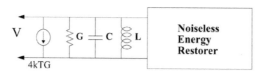

- Can show that the noise-to-signal ratio is

$$\frac{N}{S} = \frac{\overline{V_n^2}}{V_{sig}^2} = \frac{kT}{E_{stored}} = \frac{\omega kT}{QP_{diss}}.$$

- Noise current then sees pure LC impedance.

Let's look at a lossy LC oscillator; as it is shown the network can be modeled by an RLC which R represent the loss and the noise is modeled by a current noise. To have the oscillation there should be an energy resonator in parallel with LC load. It can be shown that the signal to noise ratio is dependent to the dissipated power and the quality factor (Q) of the resonator. For the oscillation to happen, the negative resonance generated by the energy resonator must cancel out the loss of the tank in the steady state condition. Therefore the noise current sees lossless LC impedance.

Phase noise: General considerations

  8

- ## Practical oscillators operate in one of two regimes:
  - *Current-limited*, where oscillation amplitude is proportional to $I_{bias}R_{tank}$.
  - *Voltage-limited*, where oscillation amplitude is independent of I_{bias}.

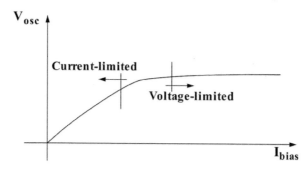

Let's consider that the energy resonator section has used a bias current of Ibias for the operation. In practical oscillator the oscillator operates in one of the two following regions: current limited region or voltage limited region. In a voltage-limited region, the oscillator amplitude is independent of the Ibias. In the other words there are some part of the Ibias, which won't contribute to the oscillator to properly work. On the other hand in the current limited region, the oscillator swing is proportional to the bias current.

9   **General considerations**

- ### In the voltage-limited regime, increases in I_{bias} do not increase carrier power.
  - Additional bias current only increases dissipation and noise, so CNR degrades.
- ### In the current-limited regime, increases in I_{bias} increase signal power faster than noise power.
  - CNR increases until incipient voltage-limiting occurs.
- ### Best CNR occurs near boundary between limiting regimes.

In a voltage limited regime, the additional bias current generates more noise and increase the power dissipation which would degrade the carrier to noise (CNR) ratio. On the other hand when the oscillator operates in current limited region, the CNR can be improved by higher bias current until the oscillator enters the voltage limited regimes. The best CNR can be met at the border of two regimes.

10   **Oscillator phase noise**

- ### The expression for CNR reveals important optimization objectives:
  - $v_{carrier} \propto I_{bias}R_{tank} \Rightarrow P_{carrier} \propto (I_{bias})^2 R_{tank}$ in the current-limited regime.
  - $P_{noise} = kT/C = kT\omega^2 L$, if dominated by tank loss.
  - So $N/C \propto kT\omega^2 L/(I_{bias})^2 R_{tank}$ to an approximation.
- ### Generally want to *minimize L/R* to optimize for a given frequency and power consumption.
  - This advice contradicts published recommendations to *maximize L*.
- ### *N/C* is important, but so is detailed spectrum.

Let's drive an equation for CNR considering an LC oscillator. As it is shown L over R ratio is a key component to be considered.

Naive LTI model

11

- Assuming all noise comes from tank loss, PSD of tank voltage is approximately

$$\frac{\overline{v_n^2}}{\Delta f} = \frac{\overline{i_n^2}}{\Delta f}|Z|^2 = 4kTG\left(\frac{1}{G}\frac{\omega_0}{2Q\Delta\omega}\right)^2 = 4kTR\left(\frac{\omega_0}{2Q\Delta\omega}\right)^2$$

- In equilibrium, noise power splits evenly between phase and amplitude domains. Then

$$\mathcal{L}(\Delta\omega) = 10\log\left[\frac{\overline{v_n^2}/\Delta f}{v_{sig}^2}\right] = 10\log\left[\frac{2kT}{P_{sig}}\left(\frac{\omega_0}{2Q\Delta\omega}\right)^2\right]$$

- Funny units: dBc/Hz at a certain offset freq., e.g., "-110dBc/Hz@600kHz offset from 1.8GHz."

For the phase noise model, let's start with the Naive LTI model and assume that the tank noise is the only noise source to be considered. The power spectral density (PSD) of the tank voltage can be approximated by the equation shown. Assuming that the noise power is split evenly between amplitude and phase noise, the phase noise of the oscillator can be formulated as below.

Naive LTI model vs. reality

12

- Previous expression doesn't describe real PN well:

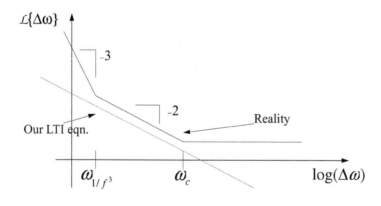

Comparing the results of the Naive LTI model with the real model of phase noise, shown in some slides back, is not promising; the comparison shows that the model can't predict the different regions of the phase noise spectrum.

13 Leeson model

- Leeson famously provided heuristic fix to remove discrepancies:

$$\mathcal{L}(\Delta\omega) = 10\log\left[\frac{2FkT}{P_{sig}}\left\{1 + \left(\frac{\omega_0}{2Q\Delta\omega}\right)^2\right\}\left(1 + \frac{\Delta\omega_{1/f^3}}{|\Delta\omega|}\right)\right]$$

- F accounts for excess noise at all offsets.
- $\Delta\omega_{1/f^3}$ accounts for $1/f^3$ region near carrier.
- First additive factor of 1 accounts for noise floor.
- These factors cannot be determined *a priori*; they are largely *a posteriori* fitting parameters.
- Need to revisit assumption that oscillators are LTI.

However referring to the Lesson equation, it turns out that it could remove the discrepancies. Let's look at the formula carefully and see each parameter. As we see, considering an oscillator as a LTI system is not a valid assumption.

14 Are oscillators LTI?

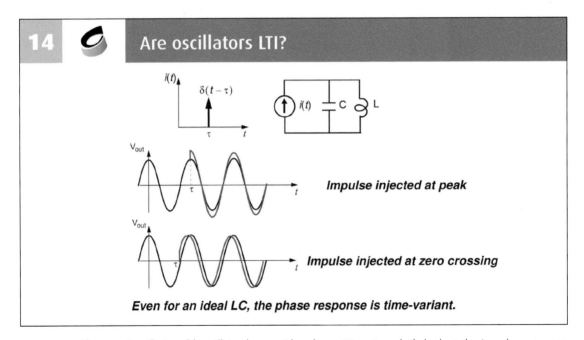

Impulse injected at peak

Impulse injected at zero crossing

Even for an ideal LC, the phase response is time-variant.

To answer the question that could oscillator be considered as a LTI system, let's look at the impulse response of LC tank. For this example let's compare the impulse response when the impulse is applied at two different time points; One at the peak of the oscillation and the other at the zero crossing of the oscillator. As it is shown, even for an ideal LC tank, the assumption of time invariant system is invalid.

Amplitude restoration in real oscillators   15

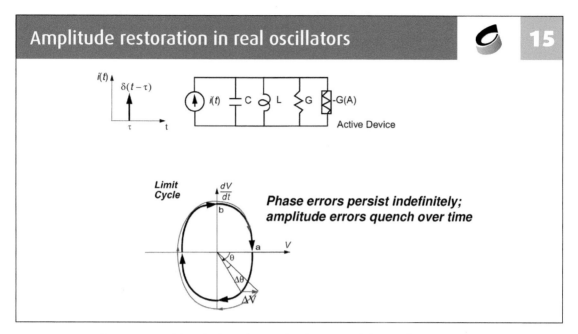

Phase errors persist indefinitely; amplitude errors quench over time

We can use the same model to look at the amplitude errors of the oscillator when an impulse tone is injected to an LC tank.

Phase impulse response   16

The phase impulse response of an oscillator is a step:

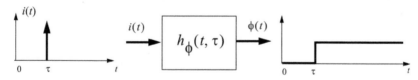

The unit impulse response is:

$$h_\phi(t, \tau) = \frac{\Gamma(\omega_0 \tau)}{q_{max}} u(t - \tau)$$

$\Gamma(x)$ is a dimensionless function, periodic in 2π, describing how much phase change results from impulse at

$$\tau = T \frac{x}{2\pi}$$

To better analysis the oscillator phase noise, let's keep working on the impulse response of the oscillator. The phase impulse response of an oscillator is a step function, which can be used to drive a time variant model for the impulse response versus the time shift for the impulse input applied to the oscillator.

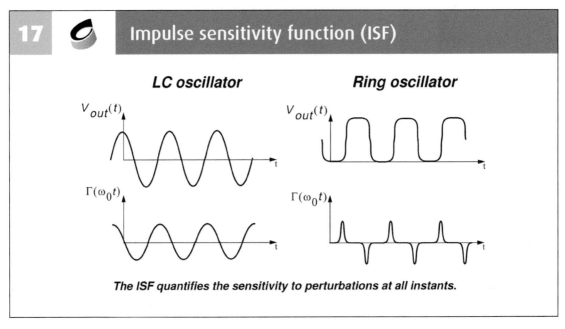

There can be defined a function called ISF (Impulse sensitivity function) which is first presented by Hajimiri. ISF can be used to quantify the sensitivity of the oscillator to perturbations at all instants. Here the ISF function is shown for two types of oscillators LC and Ring.

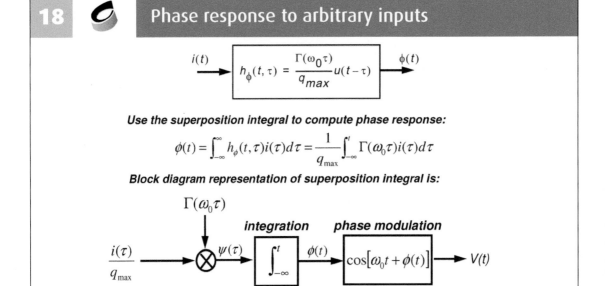

The model derived can be used to predict the oscillator phase response to any arbitrary inputs as it is shown. Use the integration and superposition to drive a block diagram representing the oscillator output voltage. The non-linear block acts as a phase modulator to modulate the oscillator output by the integrated phase noise error.

Phase noise due to white noise   19

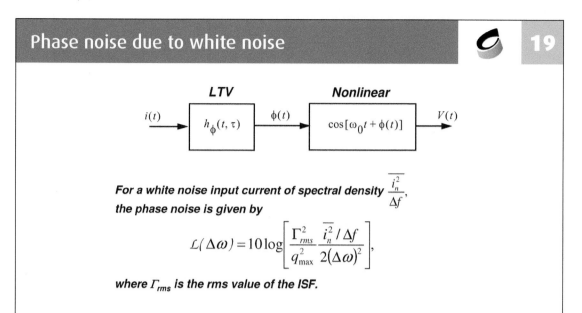

For a white noise input current of spectral density $\dfrac{\overline{i_n^2}}{\Delta f}$, the phase noise is given by

$$\mathcal{L}(\Delta\omega) = 10\log\left[\frac{\Gamma_{rms}^2}{q_{max}^2}\frac{\overline{i_n^2}/\Delta f}{2(\Delta\omega)^2}\right],$$

where Γ_{rms} is the rms value of the ISF.

The practice can be continued to evaluate the phase noise effect due to a white noise applied to the oscillator. The diagram developed for the current to voltage conversion is used, in which the phase to current conversion is modeled by a LTV system while the phase to voltage conversion is modeled by a nonlinear block.

ISF in greater detail   20

ISF is periodic, expressible as a Fourier series:

$$\Gamma(\omega_0 t) = c_0 + \sum_{1}^{\infty} c_n \cos(n\omega_0 t + \theta_n)$$

The phase is then as follows:

$$\phi(t) = \frac{1}{q_{max}}\left[c_0\int_{-\infty}^{t} i(\tau)d\tau + \sum_{n-1}^{\infty} c_n\int_{-\infty}^{t} i(\tau)\cos(n\omega_0\tau)d\tau\right]$$

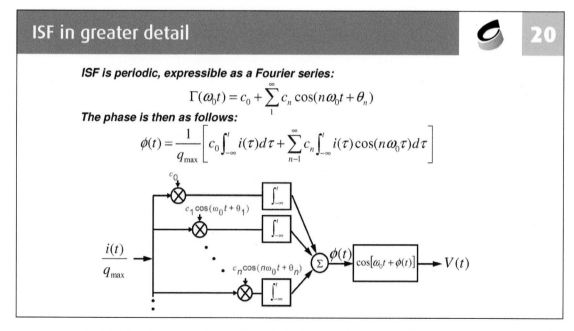

Looking back at ISF function, assuming ISF is periodic function then the implicit consequence is to use the Fourier transformation to look at all harmonic components. Following the same approach developed few slides back; the effect of each ISF harmonics on the oscillator phase noise can be calculated. The diagram below can easily show the model.

21 Contributions by noise at $n\omega_0$

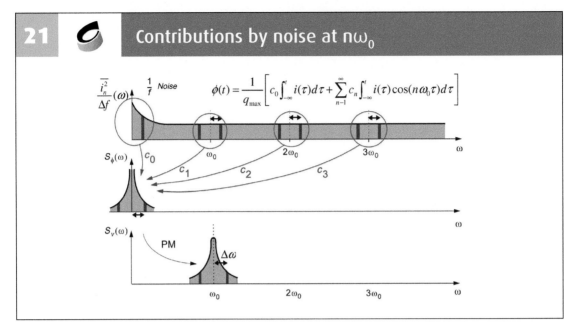

To better visualize the noise translation at the harmonics of LO frequency, let's look at the diagram shown. As the equation predicated the noise at each harmonic is transferred to the LO frequency by the corresponding coefficient.

22 Symmetry matters

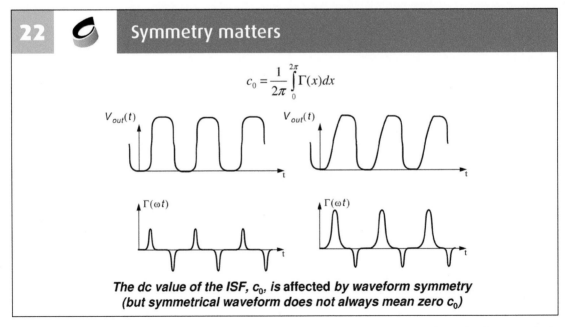

$$c_0 = \frac{1}{2\pi} \int_0^{2\pi} \Gamma(x)\,dx$$

The dc value of the ISF, c_0, is affected by waveform symmetry (but symmetrical waveform does not always mean zero c_0)

Here we take a closer look at the C0 coefficient as the dc value of ISF function. As shown by this example dc of ISF is affected by symmetry of the waveform.

1/f³ corner of phase noise spectrum

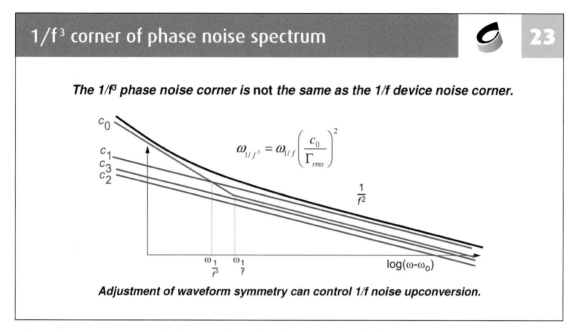

The 1/f³ phase noise corner is not the same as the 1/f device noise corner.

$$\omega_{1/f^3} = \omega_{1/f} \left(\frac{c_0}{\Gamma_{rms}} \right)^2$$

Adjustment of waveform symmetry can control 1/f noise upconversion.

The other interesting part of the ISF study is the phase noise at the frequencies very close to the carrier frequency. Let's look at the corner frequency of the $1/f^3$ region. As the equation shows the corner frequency is dependent of C_0; therefor device flicker noise up conversion can be adjusted by the waveform symmetry.

Exploring the effect of waveform symmetry

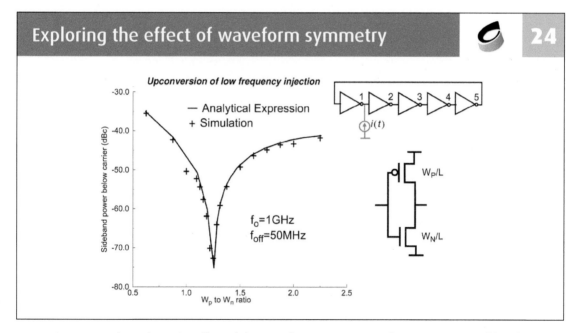

Upconversion of low frequency injection

— Analytical Expression
+ Simulation

$f_o=1GHz$
$f_{off}=50MHz$

W_P/L

W_N/L

Here is an example to show the effect of the waveform symmetry on the up conversion of low frequency injection. The example is for a simple ring oscillator made by a chain of the inverters. As it is shown there is a sweet spot in which the up converted components would be minimum.

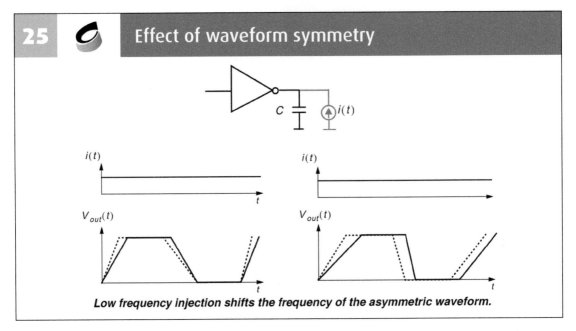

Low frequency injection shifts the frequency of the asymmetric waveform.

Here is an example showing that the frequency asymmetric waveform could be shifted injecting low frequency noise to the oscillator.

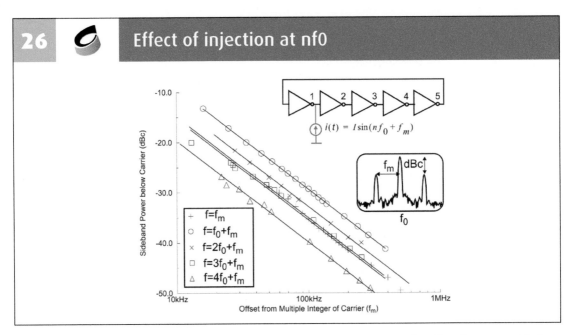

Here the experiment is done with the ring oscillator and injecting the noise at the offset frequency of f_m from the main carried or its harmonics. There would be sideband tones beside the main carrier.

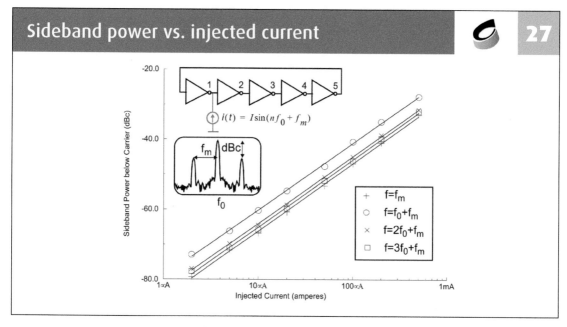

Let's look at the power of sideband tones as the injected current is increased. As it is shown the higher the injection current is, the higher the power of side band tons would be.

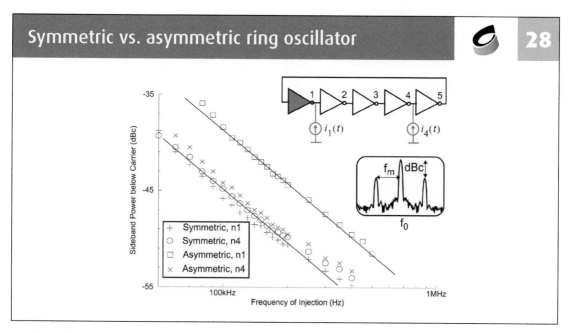

All has shown yet are based on the assumption of having the symmetric ring oscillator cells. This graph tries to compare the sideband power in symmetric versus asymmetric ring oscillator.

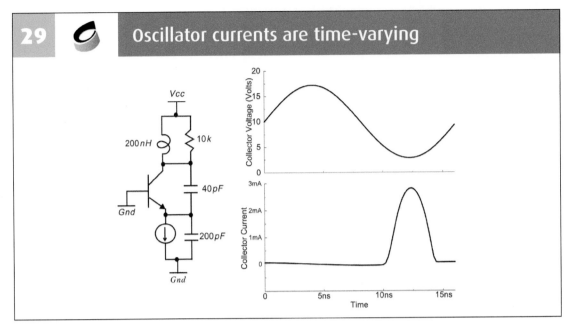

Let's look at the current of a Colpitts LC oscillator. The current is time varying signal.

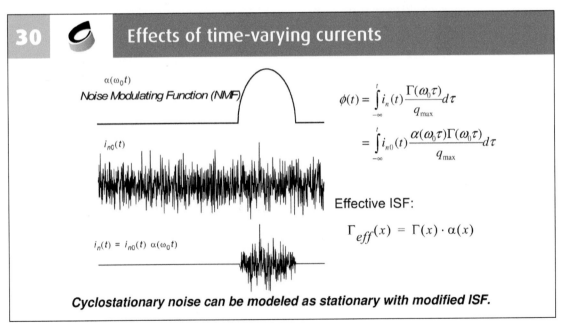

Since the current is time varying, an effective ISF function for the noise impulse response could be defined.

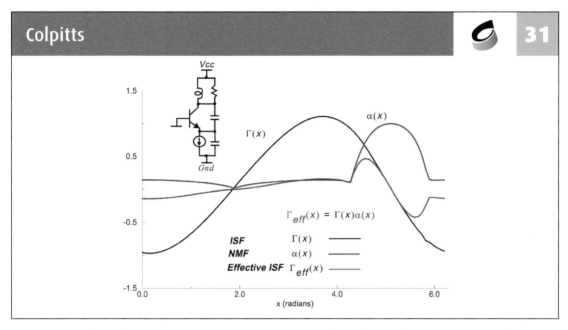

Let's look at the Colpitts oscillator one more time and now drive the ISF function versus the effective ISF function. The graph shows the current waveform as well as the ISF and effective ISF.

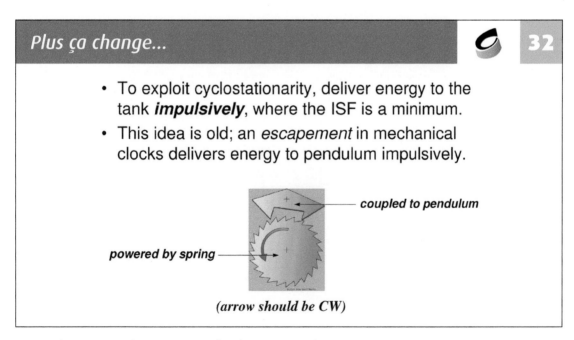

Now the question is how to improve the phase noise performance?

33 Conditions for optimal phase noise

- Impulses are delivered at or near pendulum's velocity maxima.
  - The escapement returns energy with minimal perturbation of oscillation period. [Airy, 1826]

- Similarly, the optimal moments in an *LC* oscillator are near the tank voltage maxima.

As shown by the pendulum example the optimal condition for an LC oscillator phase noise performance is to inject at the tank maximum voltage.

34 A symmetric LC oscillator

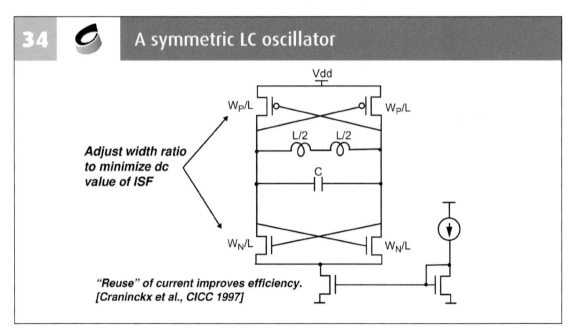

Let's look at a CMOS LC oscillator made by NMOS and PMOS cross-coupled. The architecture tries to reuse the bias current to double the cross coupled total gm. As discussed earlier, the PMOS and NMOS sizing can be done in order to minimize the dc value of ISF function, by proper adjustment of the PMOS and NMOS size ratio.

Tank voltage amplitude   35

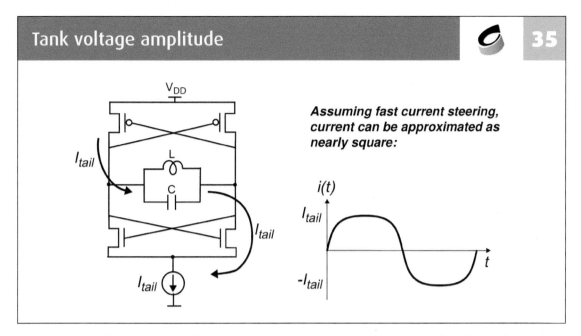

If it is assumed that switches are fast and the current steering are happening very fast, the oscillator current can be assumed to be a square wave.

Tank voltage amplitude   36

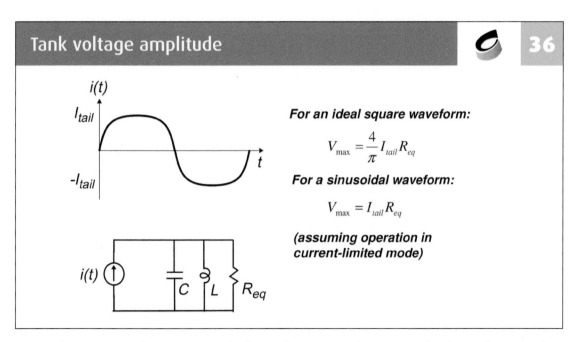

Using the square waveform assumption for the oscillator current plus assuming that the oscillator is working in the current limited regime, one can easily drive the oscillator output swing.

37 — Modes of amplitude limiting

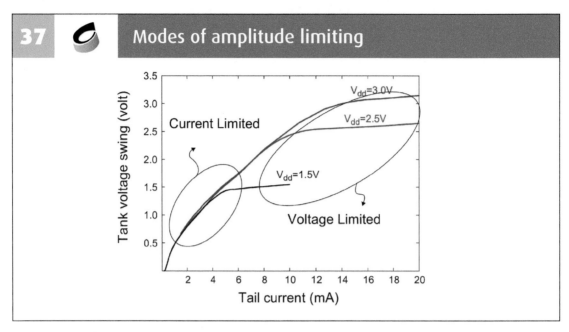

Here the graph showing the oscillator output voltage swing, as the bias current of the oscillator is swapped and the oscillator moves from the current limited regime to the voltage limited regime.

38 — Major noise sources

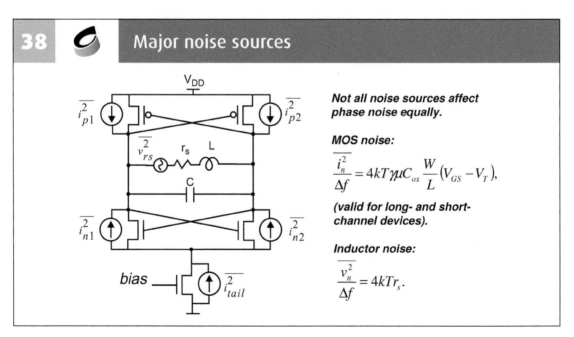

Not all noise sources affect phase noise equally.

MOS noise:

$$\overline{\frac{i_n^2}{\Delta f}} = 4kT\gamma\mu C_{ox}\frac{W}{L}(V_{GS}-V_T),$$

(valid for long- and short-channel devices).

Inductor noise:

$$\overline{\frac{v_n^2}{\Delta f}} = 4kTr_s.$$

Let's consider the CMOS LC oscillator one more time and now highlight all the noise sources in this oscillator. The question is, do all the noise sources affect the phase noise performance of the oscillator?

Equivalent circuit for noise sources   39

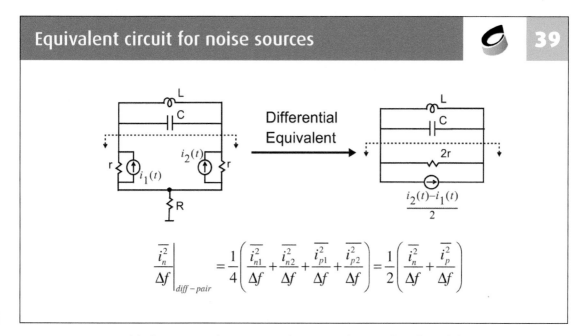

Here it tries to simplify the circuit to drive an equivalent circuit showing the effect of the noise sources.

Waveform and ISF   40

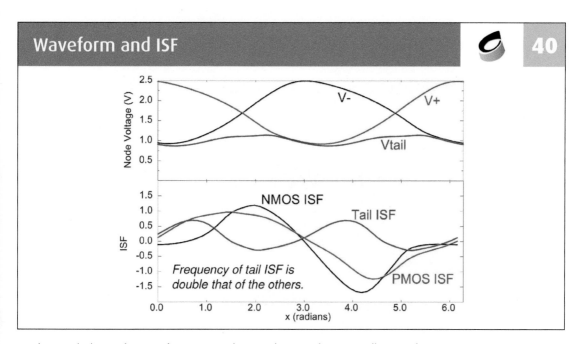

In this graph the ISF function for NMOS and PMOS devices of an LC oscillator is shown.

41 Effect of tail current noise

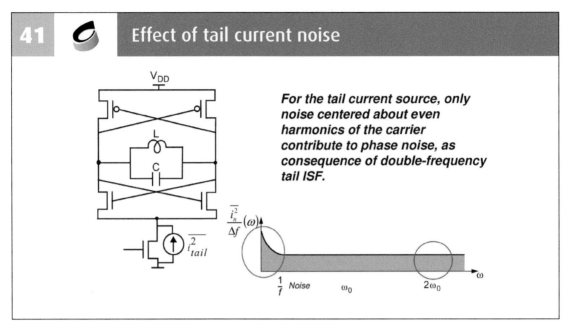

For the tail current source, only noise centered about even harmonics of the carrier contribute to phase noise, as consequence of double-frequency tail ISF.

As we saw the effect of the NMOS and PMOS noise current sources; we can now look at the effect of the oscillator tail current noise on the oscillator phase noise performance. For the tail current source, only noise centered about even harmonics of the carrier contribute to phase noise, as consequence of double-frequency tail ISF.

42 Die photo of complementary oscillator

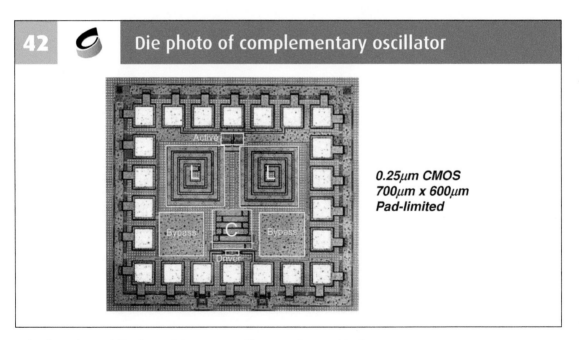

0.25µm CMOS
700µm x 600µm
Pad-limited

The floor plan and die photo of the CMOS oscillator we discussed is shown.

Measured phase noise

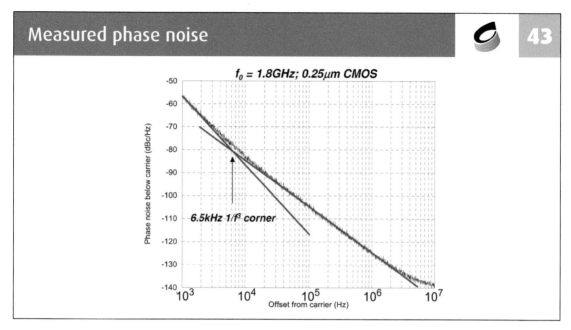

Here the phase noise measurement result is presented.

Complementary cross-coupled oscillator

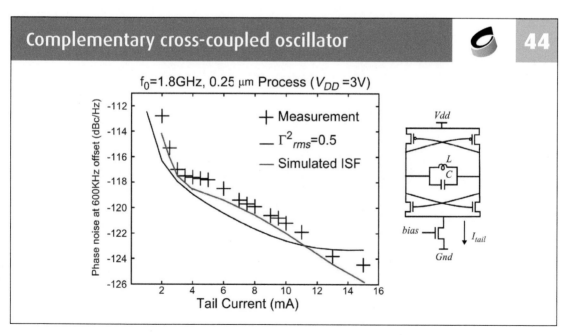

Here the oscillator phase noise performance versus the tail current is shown and the measurement and simulated results are compared together;

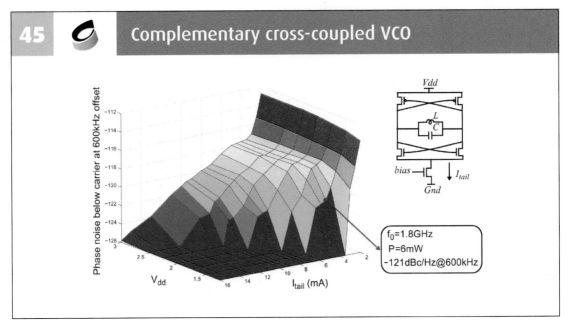

Now let's look at 3D graph showing the phase noise performance of the complementary (CMOS) oscillator for different tail current and Vdd supply voltage.

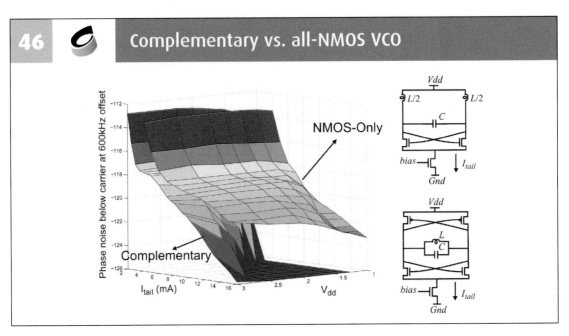

Now, let's compare the performance of two type of the LC oscillator, CMOS versus NMOS only.

A non-Stanford, non-CMOS example

47

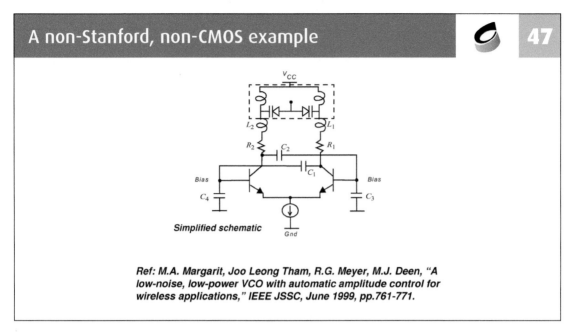

Ref: M.A. Margarit, Joo Leong Tham, R.G. Meyer, M.J. Deen, "A low-noise, low-power VCO with automatic amplitude control for wireless applications," IEEE JSSC, June 1999, pp.761-771.

All the examples till now, was based on the CMOS oscillators. Here we take a look at a BJT based oscillator with a non-standard topology.

IC and ISFs for core transistors

48

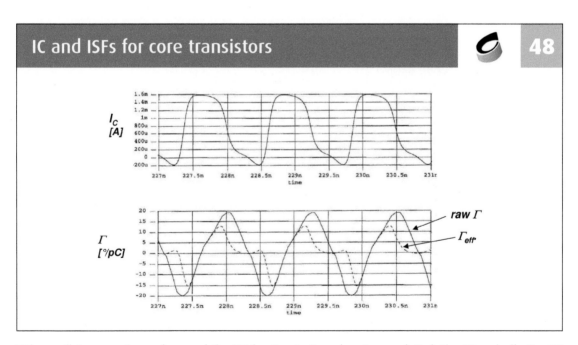

The oscillator current waveform and the ISF function in time domain are plotted. The ISF and effective ISF discussed earlier are both have been shown.

49 Vout and ISF for tail source

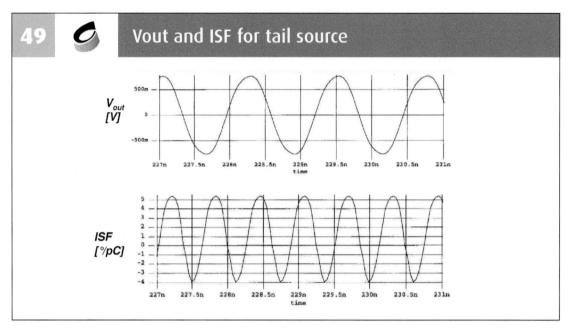

The oscillator swing and output waveform is plotted. The other waveform is showing the ISF function for the tail bias current transistors.

50 Theory vs. measurement

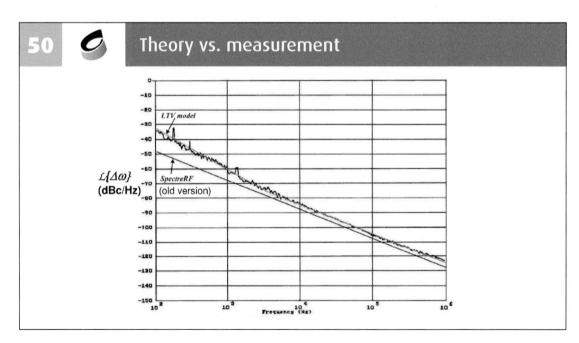

To check the phase noise performance, the graph is comparing the phase noise measurement results versus the results predicated using the LVT model.

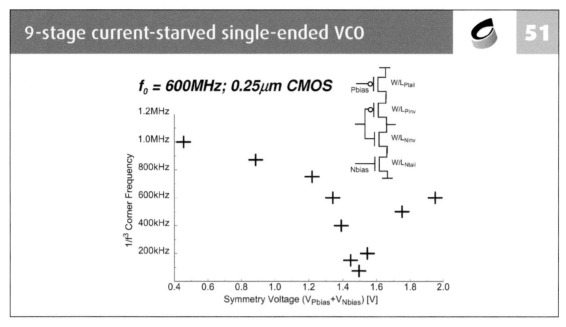

As the last example, we take a look at the ring oscillator made by chain of inverters.

Amplitude noise   52

- **Phase noise generally dominates close-in spectrum. Amplitude noise generally dominates far-out spectrum.**
- **Amplitude noise can be accommodated with the same impulse-response method.**
  - Amplitude control dynamics are frequently approximately single-pole.
    - For an isolated LC tank, the bandwidth is simply ω_0/Q.
    - Contribution to output noise spectrum is flat up to an offset equal to the amplitude-control bandwidth, then rolls off.
    - Superposition with phase noise contribution leads to a characteristic pedestal in the oscillator spectrum.
  - If dynamics are second-order, there can be peaking in the oscillator spectrum.
- **Sideband asymmetry can also result from superposition.**

All our discussion was around the oscillator phase noise; but how about the amplitude noise of the oscillator? Phase noise generally dominates close-in spectrum while amplitude noise generally dominates far-out spectrum. Amplitude noise can be accommodated with the same impulse-response method presented for the phase noise calculations.

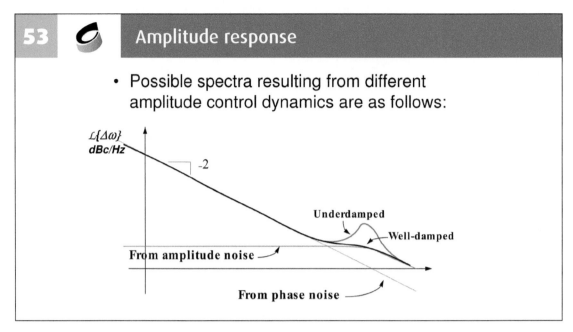

53 **Amplitude response**

- Possible spectra resulting from different amplitude control dynamics are as follows:

A s the last graph, let's look at the oscillator phase noise spectrum to highlight the regions where the phase noise would be dominated versus the regions where amplitude noise plays the main role.

54 **Summary and conclusions**

- LTI theories say:
  - Maximize signal power and Q, operate at incipient voltage-limiting, with minimum L/R.
  - One cannot do anything about $1/f$ noise upconversion.
    - $1/f^\beta$ corner is purely a function of technology and bias.
- LTV theory says:
  - LTI is generally right in the first bullet above.
  - LTI is wrong about the second bullet: Exploiting topological symmetry to minimize the ISF dc value can suppress $1/f$ noise upconversion.

H ere is the summary of the discussion to compare the results of phase noise analysis using LTI versus LTV theories. While LTI theory fails to predict the up conversion of the flicker noise in an oscillator, LTV model can suggest a recommendation to minimize the up conversion components.

References

 55

The literature on phase noise is vast. This presentation derives largely from a tiny subset:

Ali Hajimiri and Thomas H, Lee, "A General Theory of Phase Noise in Oscillators," *IEEE J. Solid-State Circuits*, Feb. 1998, vol. 33, no. 2, pp.179-194.

Thomas H. Lee and Ali Hajmiri, "Oscillator Phase Noise: A Tutorial," *IEEE J. Solid-State Circuits*, Mar. 2000, vol. 35, no. 3, pp.326-336.

Here are our two main published papers focusing on the oscillator phase noise analysis and ISF function method.

CHAPTER 2

Clock and Timing in Wireline Communications

Nikola Nedovic

NVIDIA Corp., USA

The chapter reviews the basics of timing and clock and data recovery circuits in high speed links, inluding system level considerations, limitations of the linear model, pull-in process and false lock, and a case for and key features of time domain event-driven system model to capture the effect of the circuit parameters on system performance. The second part of the presentation reviews the circuits for phase detection and modeling of samplers for fast and accurate system level modeling.

1

Outline

- **Role of the clock and types of clock architecture**
- **Clock and data recovery - system design perspective**
- **Case for event-driven time-domain link model and its key features**
- **Circuits and circuit modeling**
- **Conclusion**

In this chapter, we will review the types of the link architecture with respect to the role of the clock. Then, we will go through the system design of the clock and data recovery, specifically in plesiochronous systems. We will build a case for an event-driven time-domain link model for the purpose of design and predicting the link behavior. The rest of the presentation is going to review some of the circuits and their models that will be suitable for use in this time-domain model, and this mainly includes samplers/comparators and different types of phase and frequency detectors due to the time limitation.

2

Clocking Architectures

- **Synchronous system - Phase relationship between TX and RX clock is known (low speed interfaces)**
- **Mesochronous system - TX and RX frequencies match but the phase relationship is unknown**

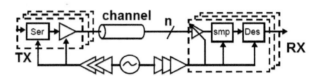

- **Plesiochronous system - TX and RX clock frequencies are slightly different, e.g. due to the tolerances in clock references**

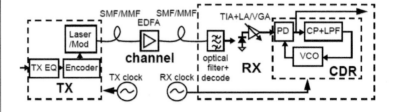

- **Asynchronous - no clock, "handshaking" needed**

IO links require some kind of synchronization between the transmitter and the receiver. We can broadly classify the clocking architectures into four types. In synchronous systems, which are used in low speed interfaces and often in on-chip links, the phase relationship between the transmitter TX clock (and therefore data) and the receiver the RX clock is known, and the RX just needs to sample the incoming data with its clock. In mesochronous systems, the TX and the RX clocks share the same frequency but the relative phase relationship is unknown, as will be detailed in the next slide. Plesiochronous systems use similar but not exactly equal frequncies, for example due to the tolerances in the clock references. Finally, asynchronous systems do not use clock, and require some kind on two-way communication between the TX and the RX to move the data.

Mesochronous System

- **TX and RX operate at the same frequency, often on the same chip or package and often on a parallel data interface**
- **Phase relationship is unknown, e.g. due to inability to control data delay in the channel or clock distribution**
- **On chip or memory interfaces**
- **Clock forwarding**
- **Need to recover the phase of the data but no need to extract clock frequency**
  - *Delay-locked loop is sufficient*

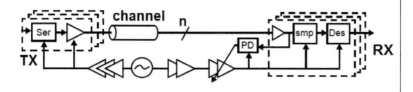

Most interesting from the point of view of modern links are mesochronous and plesiochronous systems. In a little more detail on mesochronous system, the TX and the RX operate at the same frequency, but the phase relationship between the data and the clock is unknown. This can be because we cannot control the data delay in the channel, or the clock distribution, for example the channel length variation is too large compared to bit duration. We often find these links in on-chip or memory interfaces, and they are often used in parallel links with clock forwarding, meaning that in addition to a number of data bits, the TX also sends the clock. An important note is that, since the RX knows the clock frequency, it only needs to recover the phase of the incoming data with respect to its local clock. For this, a simple delay locked loop is sufficient.

Plesiochronous System

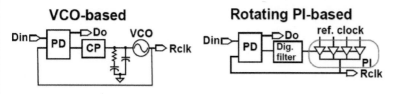

- **Must recover the clock frequency from incoming data**
- **CDR extracts clock from the incoming data stream and synchronizes the data with the clock**
- **Common in longer links (long haul, rack-to-rack, backplane) where the cost of using one lane to send the clock is too high**
- **Must use some protocol at higher communication layer to prevent data overflow/underflow**
- **PI-based CDR uses digital filter and can be viewed as an ADC**
  - *Analog input phase gets digital representation of ref. clock phase*
  - *If there is frequency mismatch, PI "rotates", i.e. outputs progressively later (or earlier) phase of ref. clock, causing Rclk to lock to Din*

Plesiochronous systems are more general, more complex, and more power hungry than mesochronous systems. In this case, we only know approximate clock frequency at best, and so the receiver must extract, or recover the clock frequency and phase, and only then sample and synchronize the data with the clock. Plesiochronous clocking is

▶

4 Plesiochronous System (cont.)

common in longer links where the clock references cannot be shared and the cost of sending the clock is too high.

It is important to note that since the TX is sending the data at the rate of its own reference or core clock, which is different than the rate at which the RX is consuming the data, there must be some protocol to prevent underflow or overflow. Normally this is done at a higher communication layer.

There are several common types of the clock recovery receivers, including all digital ADC-based, or blind oversampling, but most often we will find a PLL-based Clock and Data Recovery (CDR), a rotating phase interpolator (PI)-based CDR, or their variants. Both are negative feedback systems, where the phase difference between the incoming data and the clock is first detected and then after some signal processing, which usually involves integration and low pass filtering, a fresh clock signal is generated such that its phase follows the phase of the incoming data, which is ensured by the feedback to the phase detector. In case of PI, this "fresh" clock is generated from a set of multiple phases of the local reference clock. Although it can be analog, most PI loops are digital, meaning that the PI control, which decides the mix of the input phases that will be produced at the output, is in the form of a digital word. Therefore, the loop filter and the signal processing is normally digital, and the input phase gets a digital representation in units of some part of the reference clock phase, so this system is actually performing A/D conversion. Also note that, if there is a frequency mismatch, the PI "rotates", meaning that its control code continuously increases or decreases to make the recovered clock lock (or actually hover around) the input data, and have different frequency than the reference clock.

5 CDR Design Requirements

Here are some common specifications for a CDR. We are typically required to satisfy certain bit error rate and electrical specifications; then jitten tolerance and jitter transfer mask may be required, usually per standard. For example, jitter tolerance mask mandates that the receiver must operate correctly when sinusoidal phase jitter

- **Specifications**
  - *Bit Error Rate (BER)*
  - *Electrical specifications*
  - *Jitter Tolerance*
  - *Jitter Transfer*
  - *Alarms specifications (Loss of Lock, Loss of Signal)*
  - *Power limit*
  - *Consecutive Identical Digits (CID) limit*
- **Specific requirement for particular system**
  - *Optical system may require phase adjust*
  - *Multi-bit-per-symbol signaling (e.g. PAM-4, duobinary) may require locking to specific edges*

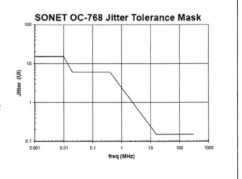

SONET OC-768 Jitter Tolerance Mask

at its input is presented on top of the data, whose amplitude and frequency are given by the mask. Often alarm specifications are set such as loss of lock or loss of signal, then power specification, maximum length of consecutive identical digits etc. In addition, some systems may have specific requirements, for example optical links will often specify phase and slice adjust.

System Design

6

System Design

The next few slides will go through a quick and simplified system design for a feedback type CDR, specifically VCO-based CDR.

Linear Model

7

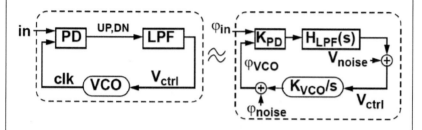

$$\beta A = K_{PD} \cdot K_{VCO} \cdot \frac{H_{LPF}(s)}{s}$$

- **Linear analysis from control systems possible and extremely useful**
  - *Stability, loop bandwidth, jitter transfer, steady-state error, noise transfer function...*
- **Linear model is only an approximation (sometimes crude)**
  - *Nonlinearities limit usability of the model, e.g. bang-bang PD*

what is its bandwidth, steady-state error, noise behavior etc. However, we must keep in mind that the linear model is only an approximation and its usefulness, and when and how it can be used, is limited by the nonlinearities, most importantly the nonlinear phase detector, as will be shown later.

Here a point is in order about the PI-based systems. While they have their own architecture and circuit design considerations, for example PI design itself, PI resolution, or clock distribution, from the system design point of view, the main differences compared to VCO-based system are a) the digital nature of the loop and the PI resolution effect on the system performance, and b) the fact that the PI does not act as an integrator, and therefore for the same system behavior, we have to add another integrator in the loop. Normally this is done in the digital loop filter. The rest of the design is very similar to the design of an analog VCO-based loop.

The linear model is probably the most useful way to understand and analyze a feedback type CDR. Our open loop gain is built out of the gains of phase detector (here linearized somehow), transfer function of the loop filter, and the clock generator, whose voltage-to-phase transfer function is that of an integrator. Once we represent our system in this way, we know how to find out if the system is stable and with what margin,

8

Other things that can be done with the linear model are the analysis and design the loop for desired noise and jitter behavior. One needs to keep in mind that, because this is a low-pass system, the input noise (or jitter) transfer function is a low-pass function, and the transfer function from various noise sources in the loop to the output (or jitter generation) is a high pass or bandpass transfer function.

Linear Model (cont'd)

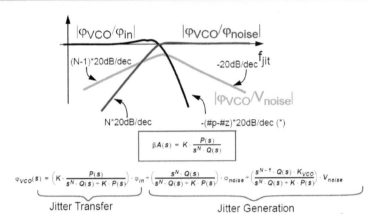

$$\beta A(s) = K \cdot \frac{P(s)}{s^N \cdot Q(s)}$$

$$\varphi_{vco}(s) = \left(K \cdot \frac{P(s)}{s^N \cdot Q(s) + K \cdot P(s)} \right) \cdot \varphi_{in} + \left(\frac{s^N \cdot Q(s)}{s^N \cdot Q(s) + K \cdot P(s)} \right) \cdot \varphi_{noise} - \left(\frac{s^{N-1} \cdot Q(s) \cdot K_{VCO}}{s^N \cdot Q(s) + K \cdot P(s)} \right) \cdot V_{noise}$$

Jitter Transfer   Jitter Generation

(*) #p-#z = difference between orders of denominator and numerator of βA

- **Output jitter has two components**
  - *Low-pass input jitter transfer*
  - *High-pass / band-pass jitter generation*

9

Loop type is determined by the number of integrators in the open loop transfer function. This is important because it determines the steady state behavior. If we apply a phase ramp (frequency step) at the input, the phase error between input and the recovered clock will be zero in steady state, in other words, there will be no systematic phase error, if the loop

Loop BW and Type

- **Loop type and numbers of poles and zeros**
  - *Normally defined by loop filter*
  - *Desired type-II (two poles at f=0) for zero steady state error for a frequency step*
    - *One 1/s term from phase integrating function of the VCO*
    - *Additional 1/s term from loop filter - typically done by an active circuit (e.g. charge pump)*
    - *In case of PI loop, we need two integrators in the loop filter*
  - *Simplest loop filter:*

$$H_{LPF}(s) = \frac{K_{CP}}{s}$$

  - *Unfortunately, with above loop filter, the system becomes unstable*
    - *Closed loop poles on imaginary axis*

type is at least two, otherwise it will depend on the open loop gain. We already have one 1/s term (integrator) from the integrating function of the VCO. We need one more integrator elsewhere in the loop, and the easiest place to do that is the active circuit that drives the loop filter (charge pump). In reality, charge pump will have some output impedance and therefore some phase error will exist, so it will be a circuit design job to minimize this effect. In case of a PI loop, two integrators will be needed in the loop filter. This is not hard to do, since the filter is digital.

The simplest loop filter is just a simple integrator. Unfortunately, with this filter, the system becomes unstable — the closed loop poles are on the imaginary axis, and in the time domain we would ideally get an oscillator in the phase domain.

Loop BW and Type (cont'd)

 10

- **Need a stabilizing zero:**

$$H_{LPF}(s) = \frac{K_{CP} \cdot (1 + sC_1R)}{s(C_1 + C_2) \cdot \left(1 + s\left(\frac{C_1C_2}{C_1 + C_2}\right)R\right)} = \frac{K_{CP} \cdot (1 + s/z)}{s(C_1 + C_2) \cdot \left(1 + \frac{s}{p}\right)}$$

- **High-frequency pole in the filter is due to parasitic capacitance at V_{CTRL}, also useful to suppress spurs**

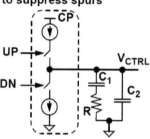

A usual solution to this problem is the addition of a stabilizing zero, and the simplest way to do that is shown in the figure. One can note that the filter also features the capacitor C2, which is responsible for the high frequency pole. This capacitor is not necessary, strictly speaking, but it is a good practice to at least model the parasitic capacitance. Besides, an actual, physical capacitor is useful to supress spurs.

Loop Bandwidth Considerations

 11

- **Loop BW↑ ⇒ input noise transfer ↑, VCO noise generation ↓**
- **BW lower bound: the knee of jitter tolerance characteristics**
- **Upper bound: defined by bandwidth of jitter transfer if specified, ultimately by loop stability**
- **Optimal bandwidth is affected by choice of reference jitter, noise of clock generator (VCO or rotating PI w/ reference clock) and other noise sources (supply, charge pump)**
  - *Generally not a good idea to position bandwidth close to a Vdd package/ PCB antiresonance frequency or large component in Vdd current profile*
- **Slight jitter peaking is unavoidable in this system**
  - *It can be reduced by lowering zero*
  - *Specified in some standards and important for repeaters due to jitter accumulation, but can be made less relevant by varying the bandwidth over the repeaters*

optimal bandwidth is affected by other factors, for example, due to the bandpass noise transfer function from the charge pump, it is not a good idea to position the bandwidth close to an antiresonance frequency of the package or PCB power distribution network.

Finally, note that in the system described here, because of the stabilizing zero, some jitter peaking is unavoidable. It can be reduced by lowering the zero frequency at the expense of the area. The maximum jitter peaking is specified in some standards, but it is important mainly in a repeater configuration to minimize jitter accumulation. Even then, it can be made less relevant by varying the repeater bandwidth.

The question now is how to decide the loop bandwidth. The tradeoff is that the higher the bandwidth, more input jitter is let through, and less (e.g. VCO) jitter is generated. The lower bound of the bandwidth is the knee of the jitter tolerance mask, and the upper bound is the jitter transfer mask, if specified, and ultimately limited by the loop stability and the VCO or the reference clock frequency. The

12

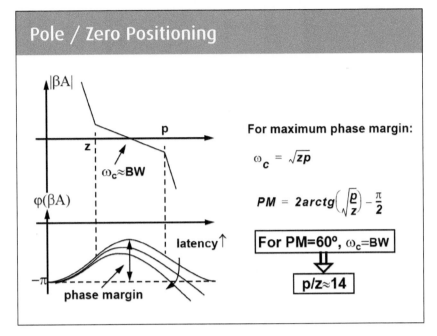

S lide 12 reviews po-
sitioning the poles
and the zero in the loop.
For the maximum phase
margin (maximum sys-
tem stability given the
pole/zero ratio), we
want to position the
open loop transfer func-
tion so that the crosso-
ver frequency (frequen-
cy at which |β A| =1) is
the geometric mean of
the zero and the dom-
inant pole. It turns out
that for the system we
are designing, if the phase margin is 60 degrees,
which is typically a satisfactory phase margin, crossover frequency is equal to the bandwidth, and it gives
us p/z=14.

13

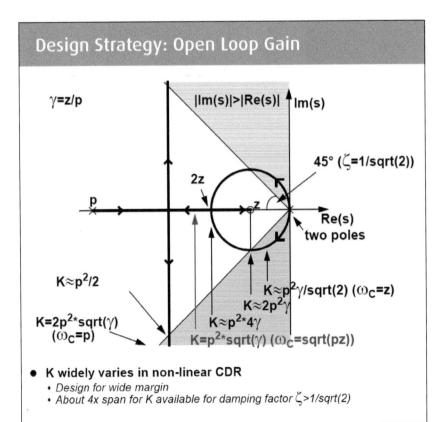

T his slide provides a
picture of how closed
loop poles move as a
function of the open loop
gain K using the root locus
method. The loop gain K
value that sets the de-
sired crossover frequency
from the previous slide is
shown in red. As it will be
shown later, because of
nonlinearities and other
effects, the desired value
of K cannot be set relia-
bly, so it is important to
allow for a wide margin.
The red design point is
also convenient in this respect, as it gives us about
2x margin for a reasona-
ble damping factor

CDR Transient Locking Behavior 14

- **Lock range is the maximum difference between VCO clock and data rate for which lock is achieved without a cycle slip**
  - All recovered data are valid
  - Useful for burst CDRs, less for continuous mode CDRs
- **Pull-in range is the maximum difference between VCO clock and data rate for which lock is eventually achieved**
  - cycle slips allowed, some data may be invalid
  - Most important quantity for conventional continuous mode CDRs
  - Plenty of literature on lock/pull-in properties of PLLs, most for first or second order loops and/or oversimplified
- **Pull-out range is maximum difference between VCO clock and data rate for which CDR stays in lock**

without the cycle slip (no data loss). Pull-in range is the maximum difference between the recovered clock and the data rate for which the lock is achieved eventually, meaning that some cycle slips are allowed. This is the most important quantity for the conventional high speed links. There is plenty of literature on

Next, the transient locking behavior of CDRs, specifically nonlinear CDRs, will be discussed. First, some definitions are reviewed, since different names have been used different references from both in the academia and the industry. Let us call the lock range the maximum difference between the recovered clock and data rate for which the lock is achieved lock and pull-in properties of PLL, and most are on first or second order loops and oversimplified. Much less literature is available on CDRs if any, and even less on nonlinear bang-bang phase detector (BBPD) CDRs. Finally, pull-out range is the maximum difference between VCO clock and data rate for which the CDR stays in lock.

What is CDR's Pull-in Range? 15

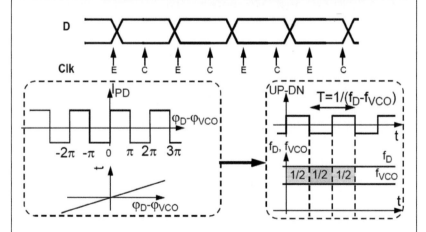

- **If VCO and data frequencies differ, VCO clock "sweeps" data eye**
- **PD output has a beat frequency component equal to f_D-f_{VCO}**
- **Area between f_D and f_{VCO} is equal to one (equivalent to one unit interval UI) over one beat period**

To answer how to estimate the CDR pull-in range, we first look at the output of a bang-bang phase detector with two signals with slightly different frequencies at its input. In this setup, the recovered clock (the VCO clock) sweeps the data eye and the PD outputs the beat frequency (f_D-f_{vco}). Note that the area between f_D and f_{vco} is equal to one (equivalent to one unit interval) over one beat period ($2*\pi$ in the phase domain).

16   CDR Lock and Pull-in

- **3rd order type-II bang-bang CDR, f_{VCO} slightly slower than f_D**

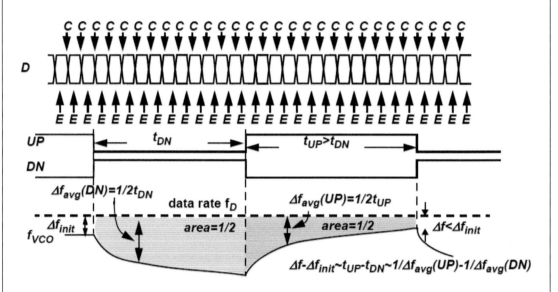

- **Pull-in happens when f_{VCO} converges towards f_D in one DN-UP cycle**

- **Change of f_{VCO} in one UP/DN cycle proportional to difference of inverses of average Δf's in UP and DN cycles.**
    - *Ideally, $\varphi(PD{\rightarrow}V_{ctl})<\pi/2 \Rightarrow$ we should always be able to pull in*

L et us now put such PD in a 3rd order type-II loop where the VCO frequency gets continuously updated. The loop is nonlinear, but because in each of the two half cycles of the beat period the output of the PD is constant, we can analyze the system in each of them in the open loop configuration. During first half of the beat period, the output of the PD is low, so the loop filter is driven by a low step and its output (and so the VCO frequency) has the smooth down ramp shape shown in the slide (without the high frequency pole in the loop filter, it would have been just a straight line). This continues until the area between the frequencies f_D-f_{vco} doesn't fill up 1/2 (half of the unit inteval or UI), at which time the PD applies the step up at the input of the filter, and then the VCO frequency smoothly ramps up as shown.

The lock happens if our recovered clock frequency (VCO frequency) ends up closer, on average, to the data frequency at the end of this cycle than it was at its beginning. It can be shown that this frequency difference is roughly proportional to the difference between UP and DOWN times. This means that if the phase across the loop filter is less than $\pi/2$, we should always be able to pull in. Incidentally, or maybe not so incidentally, this is the same requirement as for the loop stability, which would imply that any stable loop should have infinitely wide pull in range.

However, we know that this is not true in reality and that there is a lot of design effort to make sure the CDR locks, so one needs to understand what happens in a less ideal situation.

CDR Pull-in with Latency in the Loop   **17**

- **Pull-in in 3rd order type-II CDR with latency in the loop**

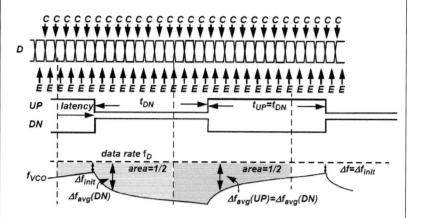

- **Pull-in is limited by latency of PD/DEMUX**

One example of a less ideal situation is a loop with latency. In this case the dynamics is the same as before, except that the input to the loop filter has a nonzero latency, which time-shifts the two half-cycles, and we can see that we can easily end up with a false lock (Δ f equal to Δ f init).

Pull-in   **18**

- **Pull-in depends on the phase of the loop gain**
  - *If $\varphi(\beta A(s=j|\omega_{VCO}-\omega_D|))=-\pi$, there is no pull-in (false lock)*
- **Other factors that affect pull in**
  - *Data transition density*
    - We can always construct a pathological (or not-so-pathological) case with more data transitions during "down" time than during "up" time so the lock never happens
    - Normally some assumption about data transition statistics is necessary. Simplest is to require minimum transition density through coding or preamble
  - *Asymmetry in PD and charge pump*
  - *Jitter and noise, especially periodic*
- **A relatively safe pull-in range is about loop bandwidth as there is a defined phase margin for stability (latency is accounted for) - x100ppm-x1000ppm**
- **With so many variables and many things that can go wrong, lock is THE most important process to model and simulate**
  - *Use as low abstraction level as possible*
  - *However, circuit level simulation is prohibitively slow*
    - Thousands of gates for millions of cycles

happens; asymmetry in the PD and the charge pump, or jitter/noise, especially periodic.

A relatively safe pull in range is about the loop bandwidth, since at around that frequency we are already designing the loop to have a defined phase margin for stability. Same phase margin works for the pull-in to happen.

The main takeaway is that there are many

In general terms, pull in depends on the phase of the loop gain. We have seen how latency reduces the pul in, and there are also other factors: variable data transition density — we can always construct an unfortunate case with more valid data transitions during down time than during up time, so the lock never

variables and things that can go wrong, so the lock is arguably the most important process to model and simulate. This simulation should be in time domain to capture all the nonlinearities and nonidealities, but it also needs to span thousands to millions of cycles. This is the problem that will be discussed next.

19

Link Modeling

Link Modeling

20

Link Modeling

- **Type of models**
  - *Behavioral high-level models - allow us to estimate stability, optimal equalization parameters etc, but are oblivious to circuit and device effects*
  - *Low level circuit-based models - accurate but too big and too slow, so a microsecond range simulation is infeasible*
- **Very hard to simulate and prove e.g. lock of the CDR with either**
- **Task: model a system to predict high-level performance while accounting for low-level effects**
- **Time domain model is preferable in order to capture behavior during the lock and effects of nonlinearities during normal operation**
- **The best chance we have:**
  - *Characterize (abstract out) low level components and effects*
  - *Use time-domain but event-driven simulation to limit number of simulation points to a single digits per unit interval (or clock cycle)*

We usually see two types of link models: high level behavioral models that help us estimate system variables — stability, statistical eye diagrams, optimal equalization etc., but cannot efficiently factor in circuit and device effects; and low level circuit models, which are accurate but too big and too slow to enable microsecond or millisecond range simulation. Neither allows us to prove the CDR lock, for example. So, our task is to model a system to predict a high level performance while accounting for low level effects somehow. The time domain model is arguably preferable so we can capture complex behavior during the lock, and the effects of nonlinearities, equalization adaptation, mutual effect of multiple loops etc.

during normal operation. The strategy that has the best chance is to characterize at least key low level components and effects — jitter, VCO/PI noise and phase offsets, sampler behavior etc. and then use a time-domain but event-driven simulation to speed up the execution.

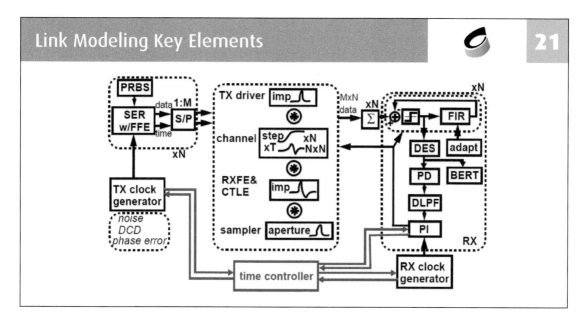

Slide 21 shows a simplified schematic of such link model. The key element is the time controller, which controls the time order of the execution. Each block that needs to be executed at a given time will send a request to the time controller, which will select the request with lowest requested execution time, send the trigger to the requesting block, which signals that the execution may proceed, and update the current time to the requested time. As a part of the execution, the triggered block is responsible for sending the next request (e.g. another data edge one UI later).

Another key feature is that all analog blocks (e.g. TX driver, channel s_{21} or crosstalk) are precharacterized with step or impulse responses and convolved together into the system response(s). Every time the RX needs to sample the channel, it actually samples the system response values at the difference between current time and times of M most recent TX or clock events, and sums them up with appropriate signs and weights to account for different types of transitions. This is the input to the digital subsystem. All blocks can then add appropriate nonidealities — jitter, offset, duty cycle distortion etc.

Link Model

  22

- **Each autonomous block (e.g. clock generators, PI, VCO) or related block issues the time of its next event to time controller**
- **Time controller picks the requesting event with smallest time associated with it, issues the trigger event to the corresponding block, and updates current time**
- **All analog blocks (TX driver, channel, CTLE, sampler) are precharacterized with their step or impulse response and convolved together into system response**
- **Every time the RX needs to sample channel, it samples and sums the system response values at the difference between current time and time of M most recent TX events. This is the input of the ideal sampler**
- **All blocks can add their specific nonidealities, e.g. noise, DCD, time domain jitter etc.**

23

Link Model

- **Relatively simple model variations for different flavors of the link**
  - *In clock forwarding architecture we send clock through a channel and write a block for estimating zero crossing at the RX*
  - *VCO-based CDR needs a VCO model, frequency acquisition loop and analog loop filter (transform loop TF to digital domain)*
  - *Different CDR rate use lower clock frequency, low rate DFE*
- **The link model code can be written in any high level language (C, perl, Matlab etc)**
  - *For fastest execution, compile and run an executable (e.g. C, Simulink)*
- **Goal is to achieve 10^4-10^6 UI/s to be able to simulate microseconds to milliseconds in reasonable simulation time**
  - *Capture lock*
  - *Estimate BER~10^{-6} to 10^{-8}*

The event-driven model can be used for different link flavors. For example, the model for a clock forwarding architecture will send the clock through the channel and include a (fictitious) zero-crossing detector block at the RX that talks to the time controller to ask for a fresh RX clock edge at the time of this detected zero crossing. A VCO-based CDR needs a VCO model (with its own links to the time controller), a frequency acquisition, and an analog filter, represented by a digital filter (e.g. by means of bilinear transform) to comply with the event-driven paradigm.

The link model code can be written in any high level language, but for the fast execution, we would like to compile and run an executable (for example C or Sim-ulink). Roughly speaking, the performance goal is to achieve tens of thousands to millions of UIs per second to be able to simulate microseconds to milliseconds in a reasonable simulation time.

The last part of the chapter deals with circuits, specifically phase and frequency detection and samplers/comparators, and their models suitable for use in the above link model.

24

Circuits

Circuits

Aided Frequency Acquisition

- **As pull-in can be guaranteed only for a few hundred ppm around data rate, phase tracking CDR's use aided frequency acquisition**
- **Most common: dual loop**

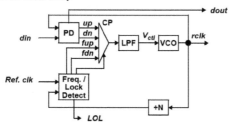

- **Frequency detector monitors the frequency difference between rclk and Ref. clk**
  - *Switch to frequency acquisition loop if greater than a threshold*
  - *Switch to phase loop if less than a threshold*

We start with frequency acquisition, in other words how to get to those several hundreds of ppm from the data rate that the phase loop needs to lock reliably. Most common frequency acquisition system is a dual loop, which switches between normal phase loop and the frequency acquisition loop and back when some monitor circuit detects that the clock frequency is close enough or too far from the data rate.

Frequency Detection Choices

- **To sense the frequency difference between two signals, we must sense the *direction* of the phase difference**
- **Most common are clock-to-clock detectors**
  - *PFD (analog output, nonlinear)*
    - To win, edge must arrive first after reset. Lower frequency clock can never do that

  - *Quadricorrelator (digital output, linear)*
    - Cross-correlate the change of in-phase demodulation with Q phase demodulation

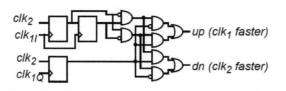

two clocks in the pulse width of its output. An interesting and a useful note is that, while this circuit is a linear phase detector, it is a nonlinear frequency detector, since only the sign of the frequency difference is encoded in the PFD, and the average of the PFD output does not change with the change of the change of frequency difference. After the reset arrives,

The key block in these loops are frequency detectors (FDs). FDs typically work by sensing the direction of the phase difference. Most common are clock-to-clock FDs, which are easier to make since, as opposed to the data, the clock is a predictable 0101... signal. By far the most popular clock-to-clock is the phase and frequency detector (PFD), which codes the phase and frequency difference between the branch that arrives first is the one that produces significant output. A lower frequency clock can never do that.

Another FD circuit is quadricorrelator, which requires quadrature phase of one of the input clocks (let's call it master clock), and works by IQ demodulating the second clock and correlating the phases of the demodulated signals.

27

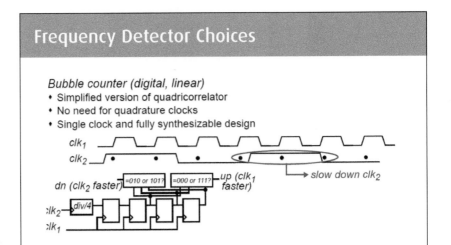

Frequency Detector Choices

Bubble counter (digital, linear)
* Simplified version of quadricorrelator
* No need for quadrature clocks
* Single clock and fully synthesizable design

A much simpler and easier to understand version of quadricorrelator is what can be called the bubble counter. It uses the single phase of the master clock, and divided-by-4 second clock. We simply look at the values of adjacent samples of the second clock. Nominally, when the two input frequencies are equal, we should see two high samples followed by two low samples and so on. A "bubble" (any different pattern) means that the master is either too slow or too fast, as shown.

28

Frequency Detector Choices

* **Clock-to-data detectors: get information at rate higher than $2f_D$**
* **Monitor the direction of the movement of the data transitions in units of the clock phase step**
  * *E.g. 3x data oversampling*
  * *No need for separate frequency acquisition loop*

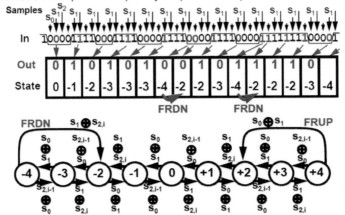

This slide shows an example of the clock-to-data frequency detector. The key to its operation is to sample the data with the clock more than twice. By comparing the adjacent samples, we can detect the direction of the phase change between the clock and data. For example if we make three data samples s_0, s_1 and s_2, and at some point we detect a transition between s_1 and s_2, then some time after that, a transition between s_2 and s_0 of the next UI, then between s_0 and s_1 and so on, we know that the clock is too fast — we have detected the frequency difference.

Frequency Detector Choices

29

- **Alexander PD**
  - *Suitable for high speed operatio*
  - *Outputs sign of the phase difference (bang-bang)*

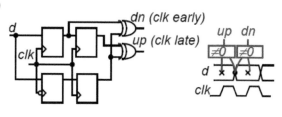

- **High Speed Alexander PD**
  - *Sample boundary sample with data s$_i$*
  - *Only if data sample experienced transition, up/dn is generated and valid*
  - *At high data rates, easier to meet timing than with Alexander PD*
    - $t_{SU} \leq t_{CLK}/2$ (as opposed to $t_{SU}+t_{CQ} \leq t_{CLK}/2$ with Alexander PD)

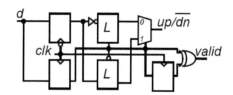

- **Hogge PD (linear)**
  - *Compare delays: data-to-rising-edge to rising-edge-to-next-falling-edge*
  - *But up/dn pulses can be very short*
    - *downstream circuit (charge pump) must be fast*
  - *Systematic phase offset equal to clock-to-output delay of first FF may be significant at high data rates*

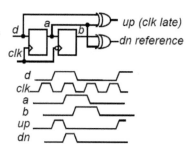

Typical textbook phase detector choices are nonlinear (bang-bang) Alexander PD and linear Hogge PD. Alexander PD is probably the most widely used PD circuit. It samples the data in the center of the data eye, and also at the boundary between two data bits. If the boundary sample is the same as the previous data and differs from the next data, the clock is fast; in the opposite case it is slow.

The slide also shows what can be considered a variant of Alexander PD, where we also make a boundary sample and a center sample, but then use the center sample to sample the boundary. Only if data experienced a transition, the up/dn is gener-

ated and valid. The benefit of such circuit is seen at very high data rates, where it is easier to meet timing than with conventional Alexander PD.

Hogge PD compares the delays: data-to-rising clock edge versus rising-to-next falling edge. If the former delay is longer, the clock is late, otherwise it is early. Ideally, the output is a linear function of the data-to-clock delay. However, the up/dn pulses can be very short and the downstream logic must be very fast to avoid nonlinearities such as dead zones. In addition, there is a small systematic phase offset equal to clock-to-output delay of the first flip-flop.

30 Effects of BBPD Nonlinearity

- **In a nonlinear system, concept of transfer function and bandwidth does not exist**
- **In reality, there will be a finite slope with gain dependent on noise, jitter, metastability, clock frequency**
- **Statistically, gain of PD can be linearized through VCO jitter and/ or input jitter**
  - *Implies the system parameters such as bandwidth, jitter transfer, peaking etc depend on noise!*

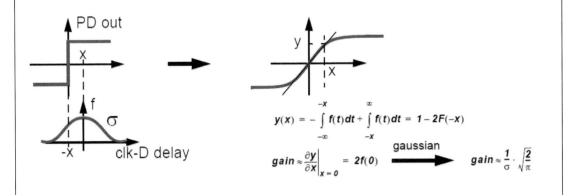

$$y(x) = - \int_{-\infty}^{-x} f(t)dt + \int_{-x}^{\infty} f(t)dt = 1 - 2F(-x)$$

$$gain \approx \left.\frac{\partial y}{\partial x}\right|_{x=0} = 2f(0) \quad \xrightarrow{\text{gaussian}} \quad gain \approx \frac{1}{\sigma} \cdot \sqrt{\frac{2}{\pi}}$$

This and the next slide go over the effects of the nonlinearity introduced by the BBPD (e.g. Alexander PD). Strictly speaking, because of such nonlinearity, the concepts of transfer function and bandwidth do not exist, and we would not know how to go about the most basic loop design. In reality, however, because the PD is normally followed by a big averaging circuit (low pass loop filter), we can characterize PD using some kind of average or statistical characteristics. For example, if our PD input or recovered clock experiences high frequency jitter, shown in slide as gaussian, we see that the output is high only in some cycles, whose probability depends on the nominal data-to-clock delay. Effectively, the statistical PD gain can be calculated and found to be a finite number. In this case this gain is a function of noise standard deviation. Note that this means that the system quantities — bandwidth, jitter transfer characteristics etc — depend on the noise!

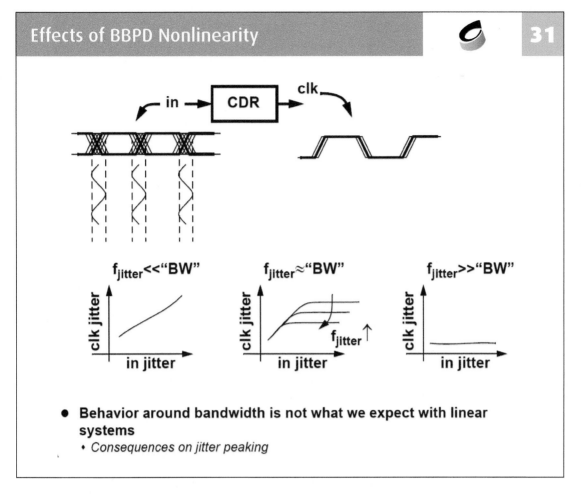

Effects of BBPD Nonlinearity   31

- **Behavior around bandwidth is not what we expect with linear systems**
  - *Consequences on jitter peaking*

Another little-studied effect of the nonlinearity is the behavior of the system (loop) around the bandwidth. If we apply an input to the CDR that has some sinusoidal phase modulation at a low frequency (below the loop bandwidth), then slowly increase its amplitude and record the amplitude of the recovered clock, we obtain what we expect from a normal linear system — the clock jitter has some linear dependence on the input jitter. If now we start increasing the jitter frequency, in a well-behaved linear system, this gain would slowly drop but would not depend on the jitter amplitude, until at very high frequencies, the gain would practically be zero and the output would not react to input. In BBPD loop case, the mode of going out of bandwidth is quite different — instead of decrease of gain, the output jitter starts saturating, and this compression point decreases as the jiter frequency increases around the "bandwidth". This behavior is essentially due to the phase slewing and has consequences on jitter peaking, but just like the bandwidth, jitter peaking should be defined differently and with jitter amplitude in mind, before a proper characterization could be made.

32   Baud-Rate Phase Detectors

- **If the channel is lossy, we need only one clock phase to sense the data-to-clock phase difference**

$$rxin_N = \sum_{k=-L}^{M} c_k d_{N-k} \qquad\qquad err_N = rxin_N - d_N = \sum_{k=-L}^{-1} c_k d_{N-k} + \sum_{k=1}^{M} c_k d_{N-k}$$

$$E(d_N \times err_{N-1}) = E\left(d_N \times \left(\sum_{k=-L}^{-1} c_k d_{N-k-1} + \sum_{k=1}^{M} c_k d_{N-k-1}\right)\right) = c_{-1} \quad \text{(assuming uncorrelated data)}$$

$$E(d_{N-1} \times err_N) = E\left(d_{N-1} \times \left(\sum_{k=-L}^{-1} c_k d_{N-k} + \sum_{k=1}^{M} c_k d_{N-k}\right)\right) = c_1 \quad \text{(assuming uncorrelated data)}$$

- **If our PD circuit uses these correlations to drive the recovered clock phase, the loop will settle at the phase where $c_{-1}=c_1$**
  - *this is known as Mueller-Muller PD*
- **Important variant is sign-sign MMPD where we process signs of d and err, and multiplication can be replaced by XOR function**
- **We can choose any other correlation of data and reference (error)**
  - *e.g. using $E(d_N \oplus err_{N-1})$ alone as UP/DN drives c_{-1} to zero*
  - *e.g. UP=$k*E(d_N \oplus err_{N-1})$, DN=$E(d_N \oplus err_{N-1})$ results in $k*c_{-1}=c_1$*

The next several slides will review a class of phase detectors that require only one clock phase per unit interval, and instead rely on lossy channel that spills some of the bit energy to the adjacent data samples (intersymbol interference, or ISI). We observe the current data bit $rxin_N$ and also the error sample, which measures the difference between the received data and ideal data d_N. If, for example, we calculate the expectation of the product of the current data bit — before any sampling — and previous (N-1) error value, and assume the data is uncorrelated, so that the expectation of two data bits i and j is the cronecker delta function of i and j, we find out that this expectation is proportional to the value of the first precursor C_{-1}. Similarly, cross-correlating the data with the next bit error yields first postcursor c_1.

If we build the PD that uses these cross-correlations to drive the CDR, the loop will settle at the point where te first precursor C_{-1} is equal to the first postcursor C_1. This particular phase detector is known as Mueller-Muller phase detector (MMPD). A practically important variant of MMPD is its sign-sign version, where we sample data and error before correlating, since multiplication now can be replaced with much simpler and more efficient XOR function.

Note that there are many other variants and PD's that can be built on this principle. For example, using $E(D_N \text{ XOR } err_{N-1})$ alone as UP/DN yields $C_{-1}=0$ as the lock condition. Similarly, scaling one of the legs in the original MMPD by a factor of k results in equally scaled lock condition.

MMPD with DFE

 33

- Another important MMPD variant reuses DFE reference samples for PD

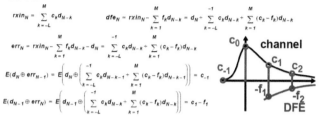

- Note that DFE loop will drive $c_1 = f_1$ so in steady state, this PD will drive c_{-1} to 0
- There is an interplay between DFE adaptation and CDR, and the loop behavior will depend on the bandwidths of the two loops
  - *If DFE adaptation loop is slower than CDR, the DFE will not be able to follow phase excursions and PD gain will still depend on c_1*

timate also includes the contribution from the DFE. Note that the DFE loop will drive $c_1 = f_1$ in steady state. Also note that there is an interplay between the DFE adaptation and the CDR, and that the loop and the DFE behavior will depend on the relative bandwidths of the two loops. If for example, the DFE adaptation loop is much slower than the

Another important variant uses PD in combination with a decision feedback equalizer (DFE), where error samples include the contribution from the DFE feedback. In this case, any postcursor es-

CDR, f_1 will not be able to follow the phase excursions and high frequency PD gain will still depend on c_1, and not on f_1.

Baud Rate PD - Pattern Filtering

 34

- Another type of baud rate PD relies on using only convenient data transition patterns
- If the data arriving to the sampler is contaminated with intersymbol interference (ISI), we can find both data patterns (e.g. 011 and 100) and the voltage offset of the PD sampler ('o' sampler) such that its output directly indicates phase error between clock and data

- Voltage offset is clearly a function of the ISI
  - *with duobinary signaling (1-D partial response), voltage offset is zero*
  - *If the postcursor equals 1/3 of the cursor ($c_1 = c_0/3$), voltage offset for PD is exactly opposite of the voltage offset of the data sampler*
    - *We can double the two samplers used for loop-unrolled DFE as the two PD samplers [Shibasaki2014]*

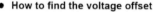

- How to find the voltage offset
  - *Use partial response signaling, thus forcing the known voltage offset*
  - *Sense the ISI (e.g. using LMS) and estimate the voltage offset during or after the CDR lock*

clock and data.

The voltage offset is clearly a function of the ISI. For example, with duobinary signaling, (postcursor equal to the main cursor), the optimal PD offset is zero. In another important example, if the postcursor is 1/3 of the main cursor, the voltage offset is exactly negative of the data sampler voltage offset. This allows us to

Another type of baud rate PD relies on using only convenient data transition patterns for purpose of phase detection. It also relies on non-negligible ISI. For example, a 1-tap postcursor eye diagram at the input of the sampler will have the optimal clock phase and data offset ('x') indicated in the figures above. We can now select the data patterns 011 and 100 and also the PD sampler offset ('o') that, sampled with the same clock phase, directly indicates phase error between

double the function of the two data samplers used in a speculative DFE configuration as the PD samplers for the opposite DFE history.

One way for determining the PD offset voltage is to use known (partial response) signaling, meaning that the offset voltage can be calculated and it doesn't change during the operation. Another option is to estimate the voltage offset from the estimated ISI, e.g. using LMS algorithm while the loop is operating.

35

The rest of the chapter will be devoted to the problem of the sampler characterization, more specifically — characterization that can be used in a system level behavioral model such as the one described in the earlier part of this chapter. Commonly, the samplers are characterized with their aperture, which is a weighting function over time that shows the relative contribution of the input at that time instant to the overall sampler decision. The aperture of an ideal sampler is a delta function, meaning that the decision depends on the data value at only one time instant. The simplest aperture-based model sees the sampler as the series of the analog front-

Sampler Characterization

- **We characterize samplers with aperture, which is a weighting function over time with which input at that time instant contributes to the decision at each clock edge**
  - *Ideally aperture is a delta function, meaning that sampler decides output based solely on the value of the input at one time instant*

$$q(T_x) = sgn\left(\int_{-\infty}^{T_x} d(\tau)A(\tau)d\tau\right)$$

- **Sampler is thus modeled with an analog front-end with bandwidth limitation, and "digital" back-end**
  - *Analog front-end is akin to a filter*
  - *Aperture is just time-inverted impulse response of an analog block*
- **Model can then be refined by including regeneration period, offset, history effect etc.**

end described with the aperture, and an ideal slicer. Looking at the definition of the aperture, we see that it can be interpreted as the time-inverted impulse response of an analog filter. A more refined model of the sampler may include other effects such as offset, regeneration, history effect etc.

36

Another important characteristic of a sampler, especially at high frequency of operation, is its history, which describes the dependence of the sampler behavior on its previous decisions. This is a low bandwidth phenomenon, coming from causes such as incomplete reset. It can be viewed as dynamic sampler offset, and in practice no more than one to two bits of history are non-negligible.

To model history, we can revise the aperture model to cover the data in the current clock period,

Sampler History

- **History effect describes different behavior of the sampler depending on its previous decision(s)**
  - *It can be viewed as dynamic sampler offset*
  - *In practice one to two bits of history are typically non-negligible*
- **Model**
  - *Extend the sampler model to explicitly include history effect*

$$q_A(T_{x,N}) = \int_{T_{x,N-1}}^{T_{x,N}} d(\tau)A(\tau)d\tau + \sum_{k=1}^{\infty} H_k q_{N-k}$$

  - *Aperture is effectively limited to the most recent clock period and everyting before that is lumped in sampler history, characterized by H_k*
- **Sampler history models an essentially low-bandwidth phenomenon**
- **No surprise that we can model it with a discrete filter**

only while everything before that is lumped into history and modeled as a discrete filter.

Aperture and History Extraction

37

- **Straightforward method - apply a convenient input waveform to DUT to expose aperture at one time point**
- **Due to highly nonlinear nature of the sampler, we can only know "analog" intermediate value q_A when sampler is metastable**

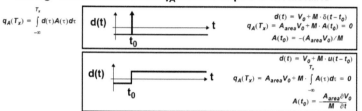

- **For each t_0, we must find appropriate $V_0(t_0)$ for which the sampler is metastable from which we can calculate $A(t_0)$**
- **Step input tends to behave nicer than Dirac input**
  - *Much easier to generate*
  - *Less numerical issues*

zero (metastability) because everything below it resolves as zero (-1) and everything above it resolves as 1.

A common approach is to apply a combination of the DC offset V_0 and either a Dirac or a step function to the input of the sampler at time t_0, and adjust the value of the DC offset V_0 to achieve metastability. According to the aperture model, the value

To extract the aperture from an arbitrary sampler netlist, we want to apply a convenient input waveform to the device under test (DUT) that exposes the aperture at one time point. However, due to nonlinear nature of the sampler, the only observable value of the signal before the ideal slicer is

of the aperture at time t_0 is directly proportional to V_0 (in case of the Dirac input) or its derivative with respect to time in case of the step input.

In practice, using the step function is better than the Dirac function as it is easier to generate, and it results in much less numerical issues.

History Extraction

38

- **Straightforward approach: extract aperture for each history**
  - *e.g. [0,0], [0,1], [1,0], [1,1]*
  - *Apertures will all have a "DC offset" (in case of Dirac input), indicating that even zero input will predictably resolve*
- **Better approach**
  - *For each sampler history, apply a DC voltage V_0 at the input*
  - *Find the value of V_0 that produces metastability at the sampler output*

  - *Solve system of 2^m linear equations with m unknowns, e.g. using least squares method*

Dirac input), indicating that even zero input will predictably resolve, due to history. A better approach is to apply only a DC voltage V_0 at the input for each history, and find the value of V_0 that results in metastability. At the end, we need to solve the system of 2_m linear equations with m unknowns, where m represents the history

The history effect can be extracted by extracting entire aperture function for each possible history. The apertures will all have a DC offset (in case of the

depth, e.g. using the least squares method.

39

Aperture and History Extraction (alternative)

- **If there is only one bit of previous history, put the DUT in a relaxation oscillator configuration**
  - *Vary t_{del} and record v_{hi} and v_{lo}*
  - *If linear discrete filter history model is correct, v_{hi}-v_{lo} should not depend on t_{del} and it should be equal to the history coefficient H_1*
  - *dv_{hi}/dt_{del} (or dv_{lo}/dt_{del}) gives aperture*

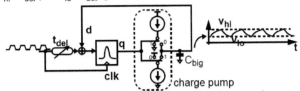

  - *Much faster simulation compared to conventional approach as we don't need to find the metastability for each d-clk time*
    - *One single (but long) simulation with slowly ramping t_{del} may be sufficient*

An alternative way of extracting the aperture and the history effect in one shot can be used if it is known that history depth m is less than two. In this method, the DUT is placed in a relaxation oscillator configuration with a charge pump in the feedback, whose output is added to the characterization step. The output will oscillate between two extremes. The difference between the two levels (the amplitude of the oscillations) is the history H$_1$ (or 2*H1, depending on the model), and the sensitivity of any of the extremes to the arrival time of the input step gives the aperture. The benefit of this approach is much shorter simulation time since we don't have to search for the metastability for each data-to-clock time point. One long simulation with slowly ramping data-to-clock time offset t$_{del}$ may be sufficient.

40

How to Use Aperture/History

- **What do aperture and history tell us about samplers?**
- **It is straightforward to use some ad-hoc metrics to compare two samplers**
  - *e.g. sampler #1 is "better" than sampler #2 if its aperture is narrower*
  - *Width: half-of-peak - to - half-of-peak, or minimum timespan for which the area under aperture is half of total area*
- **But aperture contains much more information and we can use it to predict the effect of actual data waveform on the decision**
- **According to the model we adopted, aperture function is nothing but a time-inverted impulse response of a fictitious filter in front of an ideal sampler**
  - *It can be modeled as an analog filter in front of the ideal sampler*
  - *Convolve (time-inverted) aperture with channel response, front-end CTLE etc*
- **History effect can be modeled as a "negative DFE", or a digital/discrete filter**

Aperture width is often used in a qualitative way to compare the two samplers, or to compare the aperture to the unit interval to gauge if the sampler is suitable for a given RX architecture. However, it contains much more information about how sampler will behave and we can use it to predict the effect of actual data waveform on the decision. According to the model we use, the aperture is nothing but a time-inverted impulse response of a fictitious filter in front of an ideal slicer. In our system model, the (time-inverted) aperture will be simply convolved with the other analog blocks (TX driver response, channel response, crosstalk, CTLE etc) to produce the set of system responses. History effect can be modeled as a digital filter, and its effect can be viewed as a "negative" DFE.

Limitations of Aperture Model 41

- **Poor model of sampler sensitivity and metastability**
  - *We want sampler to occasionally not resolve to hard 0 or 1*
  - *It can be improved by expanding the aperture model to include the regenerative phase in which the voltage determined by the aperture is exponentially increased over fixed amount of time [Toifl ISSCC2011 (F6)]*
- **Aperture model is as accurate as the assumption that the "analog" front-end of the sampler is linear**
  - *The assumption eventually breaks for any real circuit as we increase the input amplitude but for most useful circuits, the compression point is sufficiently high (e.g. close to maximum data amplitude or supply voltage)*
  - *Very inaccurate for single-ended static CMOS flip-flops and latches*
    - indicated by the wildly varying shape of the aperture characterized with different size steps, and d-clk dependent history

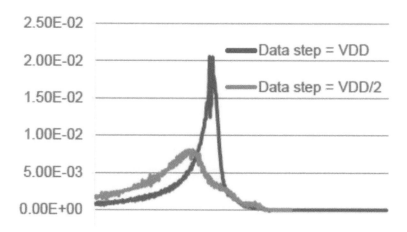

This slide briefly discusses the limitations of the aperture model. First, it does not perform very well in modeling sampler sensitivity and metastability. In reality, the sampler will sometimes output a signal that is not a hard 1 or 0, and one would like to be able to account for these effects at the system level. It is possible to expand the model with a regenerative phase in which the voltage determined by the aperture is allowed to regenerate over a fixed amount of time to help better model these effects.

More importantly, note that the aperture model is at best as good as the assumption that the data values at different time instances linearly add to act on the sampler, in other words that the analog front end is linear. This assumption eventually breaks for large enough input amplitudes for any circuit, but for most useful circuits (differential pair front-ends) this compression point is usually sufficiently high. Note that the assumption is wrong and therefore aperture can be quite inaccurate for typical single-ended CMOS static latches and flip-flops, like transmission-gate latch. This is illustrated in the slide by the widely varying shape of the aperture for different amplitude of the input characterization step.

42   Setup/Hold Time and Metastability

- **Setup/Hold times are intuitively "embedded" into aperture function, more specifically in the time offset of the aperture**
- **In a conventional 2x oversampling system, the loop settles at the median of the aperture function of its boundary sampler(s)**
  - *half of boundary samples match the previous data bit, the other half match the next data bit*
- **If we use same samplers as the data samplers, it samples furthest away from either setup/hold window, regardless of setup/hold**
- **However, this means that the boundary sampler is always close to metastability. Doesn't that cause problems?**
  - *Metastability is a concern if the signal "forks", i.e. if it is used more than once as each of downstream circuits can interpret the metastable signal differently, which may bring the system to an illegal or unanticipated state*
  - *We note that boundary sample is used only twice: once to check if the boundary equals previous data, once to check if it equals next data.*
  - *The only net effect of a metastable boundary is reduction of the effective PD gain due to occasional zero output (up=dn=0 or up=dn=1)*
  - *Even if we deserialize boundary sample prior to PD, we still "send" a single sample to only one phase detector*

Finally, this slide lists a few items that often cause confusion about samplers and CDR behavior. First, do setup and hold times somehow affect the loop settling point and do we care if they are large? The setup and hold time are somehow embedded in the aperture function, namely in its time offset. In a conventional 2x oversampling system, the loop settles at the point of the median of the aperture function of the boundary sampler, so that half of the boundary samples match the previous bit and half the next data bit. If the setup time or the hold time is large, the phase between the clock and the boundary data will not match, and the data phase of the clock will be offset from the middle of the data eye. However, if the same samplers are used for data and boundary samples, the effective point at which the clock "picks" the data will be offset by the same amount, and it is settled at a point in data eye furthest away from its setup/hold window. So, the only potential problem arises due to the mismatch in the setup/hold windows of the data and boundary samplers. Although this can be regarded as a second order effect, this mismatch will typically depend on the absolute value of the setup and hold times, so one should be careful with excessive setup/hold times.

Next, since the boundary sampler is driven by the loop to the point of metastability, does that cause problems? To answer that, we note that in general, the metastability is the problem if the signal "forks" or fans out, so that each of the downstream logic blocks can interpret the undefined metastable signal differently, which may bring the system in an illegal or unanticipated state. We do use the boundary information twice — once to check if it is different than the previous data bit, second time to check if it is different than the next data bit. So if these two blocks interpret an undefined boundary differently, the only illegal outputs we can get are up=dn=0 or up=dn=1, or something in between. This does not cause any issues functionally, but it does reduce the effective gain of the PD, which will be seen at the system level. This will hold even if we wait with PD until after the deserialization. While not catastrophic, this once again illustrates the need for use of a accurate component model in a high level system simulation.

Summary 43

- **Timing is one of key aspects of high speed link design**
- **Links typically employ complex feedback systems that span device, circuit, architecture and system design with many non-idealities and non-linearities**
- **A system-level modeling capable of capturing all design levels and execution speed capable to reveal macro level performance is necessary**
- **Many circuit blocks require careful deign and modeling**
  - *VCOs and phase interpolators*
  - *DLLs*
  - *Ser/Des at high speed*
  - *Clock distribution*
  - *Charge pumps at high frequency*
  - *Mutual dependence of clock recovery and DFE*
  - *...*
- **Phase detectors and samplers are arguably the most sensitive**
  - *Source of a coarse nonlinearity that severely affects the loop behavior*
  - *Must be accurately modeled in a way that allows for fast system simulation*

In summary, this chapter covers some topics in timing in high speed links. The key message is that the links are very complex systems that span many design levels, which interact with each other. Therefore, a system-level model capable of incorporating all or at least most critical low level effects is necessary for a predictably successful design. There are many aspects that are not addressed, such as VCO, PIs, DLL, high speed SerDes, clock generation and distribution etc Iinstead the focus is on samplers/comparators and phase detectors as arguably most sensitive circuit from the system performance perspective, as they can be a source of a severe nonlinearity that significantly affects the system performance. The point is made that these samplers must be modeled very accurately, and some ways or directions on how it can be done are provided.

CHAPTER 3

Phase Lock Loop Design Techniques

Bhupendra K. Ahuja

Qualcomm Inc., USA

This tutorial provides an overview of PLL design techniques to an IC designer. Along with basic feedback loop theory and common circuit implementations, the tutorial provides details on Integer/Frac-N, CMOS delay cell ring and LC VCO base PLL designs. Some details on current trends in Dual Loop control and Digital PLL will be provided. Details on calibration and on-chip performance monitoring will also be addressed. Jitter analysis and measurement techniques will be described.

Overview

- Introduction
- Fundamentals of PLL
- Functional Block Designs
- Jitter Analysis and measurements
- Frac-N and DPLL Designs
- Summary

The design and analysis of PLLs have been extensively studied, and published for many applications over last two decades.

This tutorial is aimed at walking thru these advances and pointing out some key design choices and criterion from these experiences.

We will briefly look at the fundamentals of PLL models followed by some design examples of more recent PLL functional blocks.

Will talk about how to estimate jitter during design and how to optimize the design for low jitter.

The recent advances in Frac-N and DPLL designs are discussed followed by a summary of the tutorial.

Applications/Metrics of PLL

Applications:
- Frequency Synthesis (Generating a 5GHz clock from 50MHz reference)
- Phase alignment and De-skewing
- Clock recovery from serial data

Performance Metrics:
- Jitter
- Frequency range and precision
- Lock time
- PVT sensitivity

Among many applications of PLL, Frequency synthesis is the primary one where a lower frequency reference clock from a crystal is multiplied up into Giga Hertz or other higher frequencies for system operation.

Other applications involve phase alignment or de-skewing of internal clock nodes to one single reference clock edge.

Serial Data communication designs with embedded clocks or source synchronous systems use PLL for transmit clock as well in clock and data recovery for the receive channels.

Each application has very specific timing requirements from the PLL or the CDR blocks.

Most critical specification is that of Jitter of the synthesized or recovered clock.

Another important one is wide frequency range of operation to meet various modes and standards.

Some applications include fast response for frequency changes and most demand robust operation over Process, voltage and temperature corners.

Functional Diagram of Charge Pump based PLL

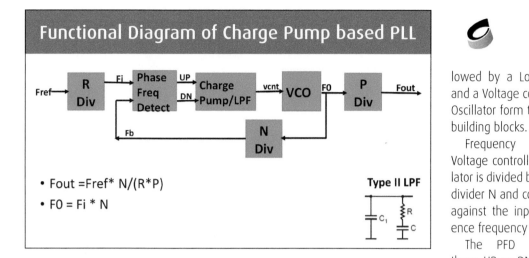

3

- Fout =Fref* N/(R*P)
- FO = Fi * N

Type II LPF

lowed by a Loop filter and a Voltage controlled Oscillator form the basic building blocks.

Frequency of the Voltage controlled oscillator is divided by the FB divider N and compared against the input reference frequency Fi.

The PFD provides these UP or DN signals

Here is quick review of charge pump based PLL design.

An input reference frequency of Fref is being synthesized here to Fout with the help of Pre, Post and Feedback dividers .

PLL is basically a second order feedback system.

A Phase/Frequency detector, a charge pump followed by

width is proportional to rising edge time differences of Fi and Fb clocks.

This Delta time information is converted first to a Charge by the Charge pump and then into a voltage by integrating this charge over a Low pass filter.

The loop filter is a series combination of R and C with a small parallel C1 capacitor, known as a Type II LPF.

Small Signal PLL Model (Wilson, JSSC,10/2000)

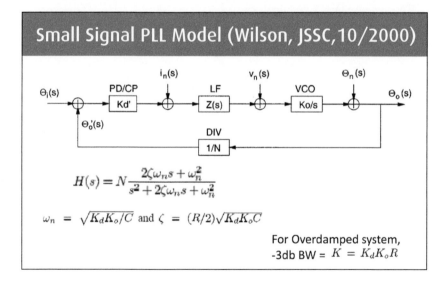

4

$$H(s) = N \frac{2\zeta\omega_n s + \omega_n^2}{s^2 + 2\zeta\omega_n s + \omega_n^2}$$

$$\omega_n = \sqrt{K_d K_o / C} \text{ and } \zeta = (R/2)\sqrt{K_d K_o C}$$

For Overdamped system,
-3db BW = $K = K_d K_o R$

The closed loop transfer function is evaluated here as a 2nd order system with complex poles and a real LHP zero.

The loop is characterized by a natural frequency, wn and damping factor rho.

The relationships for wn and rho help us design the loop for given specifications.

Here is the small signal model of the PLL, published by Wilson in 2000.

The PFD basically evaluates phase difference between the input and the feedback clock. And the ChgPump converts into a proportional charge. Thus, PFD and Charge Pump are modelled as Kd'.

The Loop filter transfer function includes a LHP zero and a pole at dc.

The VCO is modeled as an integrator.

Note that for the overdamped system, PLL loop bandwidth depends on Kd, K0 and R and Not on loop filter capacitance. We typically keep the loop overdamped due to stability requirements. Also, most systems do not like to exceed the target frequency even during any startup or mode changes.

In this small signal model, we inject noise contributions from PFD/CP, Loop Filter and VCO to study their behavior on Jitter....which we will cover later.

5

Freq-domain PLL Analysis

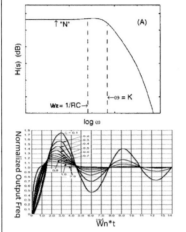

$$H(s) = N \frac{2\zeta\omega_n s + \omega_n^2}{s^2 + 2\zeta\omega_n s + \omega_n^2}$$

$$\omega_n = \sqrt{K_d K_o / C} \text{ and } \zeta = (R/2)\sqrt{K_d K_o C}$$

Wz = 1/RC $K_d = \dfrac{I_p d}{2\pi N}$

For Overdamped system,
-3db BW = $K = K_d K_o R$

The top figure shows the transfer function magnitude vs frequency plot.

Between the LHP zero and loop bandwidth K, for an over-damped system with rho > 1, it shows slight peaking near the band edge. This peaking depends on rho and determines phase margin.

The bottom figure shows step response of such a 2nd order system for different values of damping factor.

For rho < 0.707, the loop is said to be under-damped and shows long settling behavior with over and under shoots.

Frequency overshoots can be damaging to system operation due to timing margin failures.

Too much overdamped systems have long settling time and thus may not be the right choice for fast responding systems such as in disk drives.

Over PVT corners, where Charge pump currents, VCO gain, loop filter R and C vary, these relationships help us design these blocks to stay within design requirements.

6

Simple rules for PLL Stability & PVT control

: Keep Loop BW < Fref/10 with PM > 60deg.
• Keep Damping Factor > 1
• Ripple Capacitor C1 < C/20

PVT Control:
• Generate $I_p \propto V_{bg}/R$

  BW $\propto$ Ip * K0 * $\dfrac{R}{N}$ ➡ V_{bg}*K_0

  $\zeta \propto$ Sqrt(Ip*R*K_0*C) ➡Sqrt(V_{bg}*K_0*R*C)
• Make Loop Resistor R programmable/trimmable.
• PVT Invariant Self_Biased Design (Maneatis, JSSC 1996)

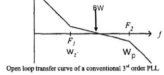

Open loop transfer curve of a conventional 3rd order PLL

The figure shows the Open loop transfer function of the PLL.

As noted, there is a double pole at DC, one due to Loop Filter and second one due to VCO phase model,

F1 is the first LHP zero formed by the Loop Filter RC time constant.

The ripple capacitor C1 introduces another higher frequency pole at wp.

For Stable operation, the BW is typically set between Wz and Fref/10.

Here are some rules of thumb for PLL Stability:

Keep PLL Overdamped over all PVT corners by choosing Damping factor >1

Keep phase margin > 60deg.

Keep higher order poles by Ripple Capacitor at 10 to 50x of Wn...this is done by making C1 < C/20.

Simple rules for PLL Stability & PVT control

For PVT control, the charge pump current is typically made by Bandgap voltage over a Poly resistor.

The Loop filter uses same Poly Resistor. For further BW control, Loop Resistor is made programmable.

Thus, the Loop BW becomes relatively constant to Vbg* Kvco.

In 1996, John Maneatis described techniques for a Self-biased PLL which keep damping factor and bandwidth to operating frequency ratio fixed over broad frequency range.

These techniques are widely used in many Analog PLL designs.

Classical Freq-domain PLL Noise Analysis

Fref/FB NTF   ChgPump NTF   Noise

$$H_i(s) = \frac{\Theta_o(s)}{i_n(s)} = \frac{\left(\frac{K_o}{C}\right)(1+sRC)}{s^2 + 2\zeta\omega_n s + \omega_n^2} \rightarrow W_n/K_d^2$$

$$H_v(s) = \frac{\Theta_o(s)}{v_n(s)} = \frac{sK_v}{s^2 + 2\zeta\omega_n s + \omega_n^2} \rightarrow K_0^2/W_n$$

$$H_\theta(s) = \frac{\Theta_o(s)}{\Theta_n(s)} = \frac{s^2}{s^2 + 2\zeta\omega_n s + \omega_n^2} \rightarrow K_\theta/W_n$$

LPF NTF   VCO NTF

Closed Loop PLL/Block Phase Noise (Rad**2/Hz)

$$\text{RMS } J_{PBR} = \frac{1}{2\pi f_c}\sqrt{\langle\theta^2(t)\rangle} = \frac{1}{2\pi f_c}\sqrt{2\int_0^\infty 10^{\frac{L(f)}{10}} df}$$

TJ = 14* Sigma + DJ

(Wilson, JSSC,10/2000)
(MAXIM, Appl Note,12/2004)

similar Low pass Xfer functions as the input.

However, the Loop Resistor thermal noise results into Bandpass characteristics while the VCO noise results into a 2nd order High pass Xfer function.

Wilson & LaxmiKumar provided a very detailed analysis of estimating jitter using these NTFs.

Important thing to note are the relationships of various sub blocks to the design parameters like wn, Ip, K0 and R.

Note that Chgpump current integrated noise is proportional to wn but inverse to Ip square.

The loop filter R integrated noise is proportional to square of VCO gain but inversely to wn.

The VCO contributions are proportional to VCO gain and inversely to wn.

The figure on the bottom right are actual simulation of Phase Noise of a LC PLL showing various contributors as per the above theory.

Since Jitter is number 1 performance metric of any PLL, it has become mandatory to estimate jitter performance of the PLL during design cycle.

Jitter is characterized as Random and Deterministic.

The random jitter comes from device noise which have low frequency noise and wideband thermal noise.

Deterministic jitter is caused by supply noise and mismatches in the design.

For a typical communication channel, TJ = 14* Sigma + DJ

Since, Random jitter gets multiplied by factor of 14, it is important to minimize it.

Looking at various noise sources and their effects on PLL jitter, above figures show noise transfer functions from input/chgpump/lpf and vco.

Note that FBclk and Charge pump Noise have

The overall jitter is computed by summing these noise contributions with their integrated NTF over frequency and then converting integrated phase noise power into jitter sigma value as shown by this equation.

8

Jitter Reduction Rules of Thumb

- Get a Clean Fref source (as high as possible)
- Increase Loop BW (W_n) and Charge Pump Current (I_{ch}) by same ratio (within Stability limits of Fref/10)
- Reduce VCO gain K_0 by using Multi-range and/or Dual Loop control.
- Reduce Impulse Sensitivity Function of VCO (fast transitions➜ PN)
- Use of LDOs for lower DJ: PFD/FB Dividers, ChgPump/LPF, VCO↓

One of the commonly used technique to reduce the PLL output jitter is to increase the loop bandwidth by increasing wn.

The only drawback with larger loop bandwidth is that more noise from the input clock will transfer to the output.

Fortunately, for most applications, input is a clean clock source with negligible phase noise.

From the previous slide noise analysis,

When wn and Ich are increased by the same factor, the phase noise transferred from Charge Pump, Loop Filter and VCO are minimized.

We also know VCO introduces significant phase noise (both random and deterministic). So, reducing VCO gain helps a lot in overall jitter reduction.

This is done by using Multi –Range and/or Dual loop control of VCO...which we will talk about in the coming slides.

In order to reduce DJ, use of LDOs has become common practice in PLL designs as we will touch on it soon.

9

Phase Freq Detector

- For PLL lock condition or very small phase difference:
  Avoid Dead-zone: UP/DN have min pulse width for ChgPump to respond

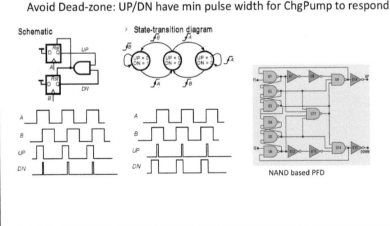

NAND based PFD

Now, let us talk about circuit design of various functional blocks.

The first one is the PFD whose design has not changed much over the years.

It is comprised of two FF triggered by input and feedback clocks and these are reset after both rising edges are detected.

A State diagram of the block is shown here along with timing waveforms which should be familiar to most practitioners.

When PLL is in lock condition, the UP DN pulses may disappear due to small gate delays. In order to keep the feedback loop active during this lock condition, one purposely makes the delay of this AND gate bit large enough to get a UP/DN pulse widths for the Chg Pump to respond.

A Nand gate based design is shown here in the bottom right.

Charge Pump with Matched Up/Dn currents

10

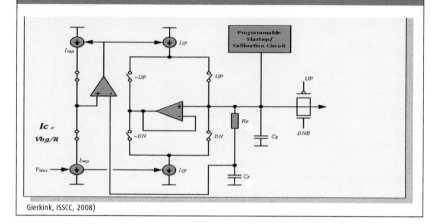

Gierkink, ISSCC, 2008)

node thru this UG buffer. This ensures value of these currents to stay same over full reference cycle.

In DSM technologies, these bias currents also depend on Vds across the transistors. For that reason, the UP and DOWN currents are made equal by having a replica leg of the output stage with turned ON switches and another feedback amplifier which forces these currents to be equal.

A Charge Pump design is shown here addressing some typical design issues.

A bias current based on Vbg over Poly Resistor is current mirrored here.

The Charge Pump currents are steered into a Dummy node during inactive phase of UP/DN pulses.

This way, the current sources do not have to turn OFF based on inactive phases of UP/DN signals .

The dummy node is basically a replica of VCNT

A set of dummy switches with opposite polarity control signals is placed here to cancel the UP/DN switching coupling on VCNT node.

Usually, there is some programmable DAC forcing VCNT upon Power up for fast startup and/or calibration purpose.

VCO/CCO Designs

11

- Current Controlled Ring of Differential Delay cells
- Voltage controlled capacitive loads inside Differential Delay cells
- LC VCO using thick metal Inductors and varactors for Low Jitter applications.
- VCO/CCO covering wide frequency range divided into overlapping ranges
    - --- Range control by digital bits
    - --- PVT calibration by Trim control bits
- Important Design parameters:
  Kvco ---needs to be constant over freq range of operation
  Low Phase Noise
  PSRR --- most design use LDO to power VCO/CCO.

they offer lower DJ due to supply/ground noise and also keep supply current mostly constant.

Most designs convert VCNT to a Current thru a V2I block and use the current to vary the delay of the cell. With inclusion of varactors in CMOS technology, one can directly control

Voltage or Current Controlled Ring Oscillators are critical design block for any PLL.

Its ability to convert a relatively slow moving control voltage into a proportional frequency signal requires careful design .

Differential architectures for RO are preferred as

Differential delay stages to applying Vcnt to these capacitive loads.

For sub psec jitter requirements, LC VCO are preferred. The on-chip inductors are formed by thick top level metal layers.

Many applications require wide frequency

VCO/CCO Designs

▶ range from few MHZ to GHz.

If one tries to design the full range in one step, it leads to large Kvco value...which we saw is not good for resulting jitter of the PLL.

So, most designers split the range in several overlapping ranges, selected by digital control bits... which may be used as Trim bits to calibrate Process variations.

Once a range is selected, the VCO/CCO should have enough frequency range thru use of VCNT to overcome any Temperature or VDD voltage varia-

tions. Thus, one can lower Kvco without compromising application requirements.

Among many design parameters, the key one is to keep Kvco nearly constant over frequency range because it effects loop dynamics.

The design should also focus on reducing random noise due to 1/f and thermal noise of the MOS transistors.

In order to reduce Deterministic jitter, most designers use an LDO to power up VCOs. We will talk more about various design techniques for VCOs next.

12

Replica Bias VCO Design

H ere is one of the first sophisticated analog design of a VCO providing a very wide PLL frequency range.

Ian Young from Intel published this replica biasing scheme back in '92 and since then this technique has been used by many ADC/ DAC/PLL/SERDES implementations.

Here a differential delay cell of the ring oscillator is shown where its delay is controlled by V2I converter.

In order to keep VCO swing constant over frequency range, a Voltage controlled resistor is used as loads in the delay stage.

A replica of the delay stage is used to generate the bias control for the VCR.

As Ictrl increases at higher frequencies, one would like to lower the VCR to keep the oscillation swing

constant. He came up with this linear + saturation region MOS combination to make the VCR.

This is accomplished this voltage feedback of the swing and comparing it against a preset value generated by this fixed swing bias block.

By keeping VCO swing constant over the frequency range, one achieves fairly linear Voltage to Frequency characteristics with tight control of Kvco.

Differential Delay Cell CCO with Multi range

 13

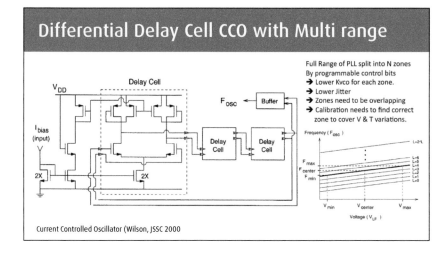

Current Controlled Oscillator (Wilson, JSSC 2000)

They introduced the concept of multi range VCO design where the full range of PLL frequency was split into multiple zones selectable by control bits.

Each Zone needs to cover now only Voltage and Temperature variations for the required frequency value.

Here is another Delay cell for the CCO design by Wilson and Laxmikumar using similar replica biasing and swing control techniques.

However, as the frequency range of PLL was approaching into GHz and VDD supply voltage was coming down from 5V to 2.5V, having a single range for PLL operation forces Kvco to be in ~400MHz/V.. thus making low jitter designs difficult.

Each Zone now can work with much lower Kvco value providing lower jitter for the overall PLL.

Zones need to be overlapping to avoid any missing range of operation.

A self-calibration loop is activated upon power up to determine the proper zone by controlling the VCO thru a frequency detect loop.

Inverters based VCO (CCO & Current Starved)

14

(Williams, CICC 2004)

means:
- current control and
- load capacitor control thru use of varactors.

Others have used pair of current controlled inverters forming a pseudo differential delay stage for the ring VCO.

The cross coupled inverters provide two functions: first to keep the two ring VCOs locked at 180deg phase offset and secondly to improve the rise/fall times of each output thus lowering ISF and jitter.

An LDO is used to provide regulated supply for these type of VCOs to lower DJ.

By 2004, the need for 2 to 5GHz VCO design in sub 2.5V rails forced designers to the use simple current starved inverters as delay cells for VCO.

A five stage VCO using Current starved inverters is implemented here by Williams.

The delay of the cell is controlled by several

15

In summary, ring based VCO are commonly used for wide tuning range applications.

Having multiple delay stages, these naturally provide multi-phase outputs which are used for Frac-N and clock and data timing recovery.

The ring VCOs also have small footprint and max out around 5GHz.

Summary of Ring based CMOS VCO

Pros:
• Wide frequency tuning range
• Multi-phase outputs: Frac-N and CDR applications
• Low Area

Cons:
• PN between -85dbc/Hz to -90dbc/Hz
• Jitter > 1ps rms
• Higher Power
• F0 < 5GHz

However, due to many devices forming the ring, its phase noise gets impacted by low freq 1/f and thermal noise.

Ring VCO exhibit PN mostly in mid to high 80dbc/Hz. Thermal noise is reduced by increasing bias currents. That is why Ring VCOs are more power hungry than LC-VCOs.

16

For many RF and >5Gbps Serdes links, the need for sub pi-co-second jitter prompts the use LC tank and a cross-coupled gain stage for the VCO.

A center tap inductor of 2 or 3 turns with an effective value under 1nh are typical range for up to 10GHz. High value of Q helps in lowering Phase Noise which dictates use of thick top

LC-VCO Designs for Lower PN/Jitter

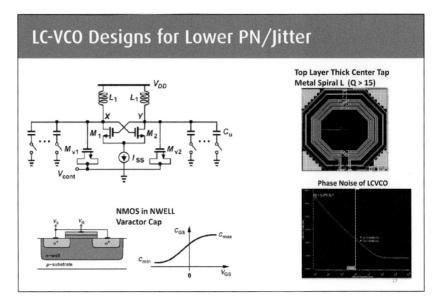

metal layers. Typical values for Q is > 15. Inductors geometries take about 50 to 100um per side area. The inductor area is generally treated as keep out area for any other signals or pads.

Most foundries provide PDK's for such Inductors for PLL and other RF applications in 65 and deeper sub nm technologies.

A programmable bank of MIM or MOM caps are used to calibrate/trim the center frequency. A bank of Varactors formed by NMOS device inside Nwell are used to provide continuous frequency tuning by the VCNT signal.

The tuning range of LC-VCO is limited to about +/-10%. For wider tuning range, multiple LC tanks are used.

Most LC VCOs use a regulated VDD. Phase Noise of such VCOs is typically above -110dBc/Hz providing rms jitter below 0.5ps.

Lower PN LC-VCO Structures

17

Max Swing: Rail to Rail

High impedance at 2 ω_o.

Hegazi and Abidi, JSSC 2001

These figures show a lower PN VCO structures.

The figure on the left maximizes swing to 2xVDD by use of cross-coupled inverters.

The figure on the right uses a tail inductor replacing the current source. Here, the tail inductor avoids the flicker/thermal noise of the MOS device resulting in lower PN.

Area of these LC VCOs is dominated by the Inductor layouts and is at least 5 times than a Ring VCO.

Feedback Dividers

18

- Critical element for PLL operation
- Must work at maximum VCO frequency over all PVT corners

Typical design tradeoffs:

- Async or Ripple counters: Usually cascade of Div 2, Jitter is additive➔ Not Done
- Synchronous Counters: Well controlled jitter but VCO freq need to be bussed ➔ power ➔ OK for Divisor power of 2 up to 256.
- Multi-modulus: Good for any count
- Frac-N PLL

Large value of N lowers the loop BW (bad for jitter) !!

Use of Async or ripple counters for N < 16 may be OK but more cascading of Div by 2 stages adds to feedback clock jitter.

Synchronous counters which use the VCO clock to trigger FF chains are good choice for Binary N up to 256. The issue here load capacitance on VCO clock buffers.

Now, let us talk about the Feedback Dividers briefly.

This is a critical element for it must work at the maximum VCO frequency. I have seen many PLL failures attributed to these dividers...so careful design and choice of architecture is warranted.

Multi-modulus dividers provide excellent solutions for any arbitrary feedback N...typically used in Frac-N designs.

Large values of N lowers the loop BW which is bad for jitter

19

Now some details on Multi-Modulus Dividers. These are used in Frac-N PLLs.

Basically we make a Divide by 2 or 3 cell shown using latches here based two control inputs C and modin.

It is a divide by 3 if both C and modin are HI. Else it is Divide by 2.

By cascading these cells as shown here, one can generate any value between two integers with step size of 1 based on the control input C.

For example for n=4, one can get any count between 8P to 8P+15.

As seen here, full VCO rate signal is applied to the first modulus divider and subsequent stages

High Speed Multi-Modulus Divider

- **Used for Frac-N PLL**
- **Only the first stage sees full VCO freq for division!!**
- **Low Power and high speed.**

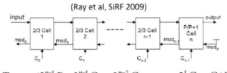

(Ray et al, SiRF 2009)

$$T_{output} = (2^{n-1}P + 2^{n-1}C_{n-1} + 2^{n-2}C_{n-2} + \ldots + 2^1 C_1 + C_0)T_{input}$$

- **For n=4, all counts between 8P and 8P+15 are possible**

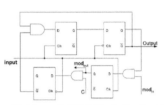

Block diagram of a divide by 2/3 cell with mod control

Div by 2 if C or mod_{in} =0
Div by 3 if C=1 and mod_{in}=1

get divided clock as inputs. Such modulators can be made to work at several GHz rates without accumulating jitter.

20

Now, let me switch gears to describe some Frac-N PLL.

Here is an example for wireless application where a reference frequency of 19.68MHz is available to generate 2402 to 2480MHz in 1MHz steps for a Bluetooth device.

If one uses integer PLL architecture, first one has to create a 40KHz reference from 19.68MHz by a pre divider of 492.

Then the feedback divider has to be programma-

Need for Frac-N PLLs: Wireless Applications (BT)

f_{ref} =40 kHz f_{vco} ~ 2.402 GHz + k MHz

÷ 492 → Phase/Freq. Detector → Charge Pump → Loop Filter → VCO → F0= 2402 to 2480MHz in 1MHz steps

19.68 MHz

÷ 60050 + 25 k

K=0 to 78 in step of 1

Impractical Integer PLL !!
(Loop BW ~ 4KHz and large FB divider)

(Ian Galton, 2003)

ble from count of 60050 to 62000 in steps of 25.

Such a design is almost impractical as it requires very low loop BW... thus, the need for Frac-N PLL came about where multi-modulus dividers are put to use.

Frac-N PLL (Wireless Applications)

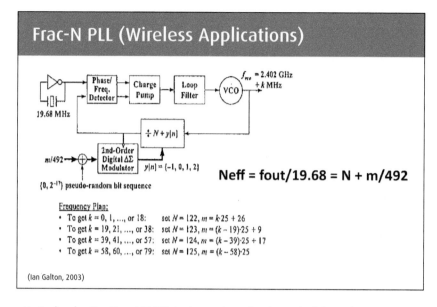

21

$$Neff = fout/19.68 = N + m/492$$

(Ian Galton, 2003)

The Delta Sigma modulator generates this m/492 fractional part by putting out a string of -1, 0, 1 or 2 as additions to integer value N.

Thus, for a given band, N is changing among 121, 122, 123 or 124 values. The delta sigma modulator makes the average value to be Neff.

A 2nd order Fractional N PLL is shown here for the same BT application here.

The feedback factor is modulated among Integer values at the feedback rate of 19.68MHz.

The average value of the feedback divider comes out to be Neff shown here.

In order to suppress buildup of quantization noise at high frequencies, the Loop filter is designed with additional poles. Fortunately, these higher order poles do not affect loop stability.

Multi-modulus dividers are preferred choice for such Frac-N applications.

ΔΣ Modulators for Frac-N PLL with Phase Interpolators

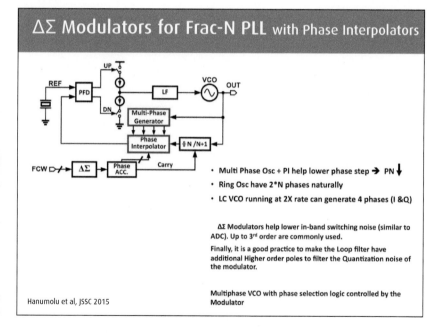

22

- Multi Phase Osc + PI help lower phase step → PN ↓
- Ring Osc have 2*N phases naturally
- LC VCO running at 2X rate can generate 4 phases (I &Q)

ΔΣ Modulators help lower in-band switching noise (similar to ADC). Up to 3rd order are commonly used.

Finally, it is a good practice to make the Loop filter have additional Higher order poles to filter the Quantization noise of the modulator.

Multiphase VCO with phase selection logic controlled by the Modulator

Hanumolu et al, JSSC 2015

This can be mitigated by making use of multi-phases of VCO.

Most ring VCO are made of 3 to 5 differential delay stages with 6 to 10 phases of VCO periods.

For low jitter, if LC VCO is used, it is run at 2X rate and one can generate 4 phases out of Divide by 2.

More recent Frac-N designs make use of this by switching between adjacent phases by the Delta Sigma modulator to create the Fractional part. The figure here shows one such implementation where available phases are further interpolated to create finer phase jumps.

One of the issue with Frac-N PLL design is reference spurs which results into increased jitter.

This happens due to the fact that feedback divider is changing between consecutive numbers and Loop filter is not sharp enough to reject this ripple on Vcnt.

23

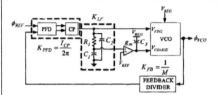

Dual Loop Control PLL

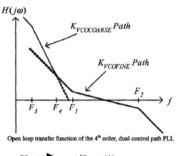

- Lowers dynamic value of Kvco to get lower Jitter
- Automatic range control thru Kvco_coarse control
- Improves dynamic range of CP by fixing Vcnt to Vref.
- For analog PLLs, bank of varactors split into two groups---one controlled by Coarse and other by Fine.

- Loop BW determined by Kvcofine.
- Concept used in DPLL designs as well !

(Williams, CICC 2004)

Open loop transfer function of the 4th order, dual control path PLL

F3 ➡ gout/C₃ << Wz
KvcoFine << Kvcocoarse (1/50)

(Typically, F3 ~ F1/25, F4 ~ F1/4)

Now let us talk about Dual Loop Control scheme for PLL...this technique has become quite frequent in modern PLL designs.

It is in some sense progression of Multi-zone VCO designs as we saw earlier in Self calibrated PLL design.

It is well understood that Chargepump produces Vcnt as sum of Proportional and Integral parts of incoming phase difference.

The integral part is the voltage across Loop filter capacitor which changes very slowly.

Williams proposed this scheme by further integrating this voltage thru a Gm-C stage to control the VCO frequency.

Thus, the VCO control is split between a coarse and fine Control inputs.

The Kv from coarse control is about 50X of Kv from the Fine control.

The coarse loop basically introduces a very low frequency pole to bring PLL frequency close to the target value and then the Fine control take over the loop.

Since Kv of fine input is much lower, the PLL jitter benefits from this.

Another significant benefit of this technique for analog PLLs is that it fixes Vcnt node to a known Vref which can be set at one half of the VDD. This helps in keeping UP/DN currents matched.

24

Use of LDOs for DJ reduction

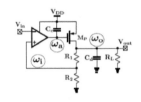

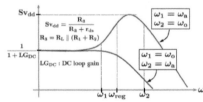

Multiple Regulators: PFD/FB Dividers, CP and VCO

To avoid High frequency peaking in PSRR: Make Output pole dominant.

Multiple LDOs, RL is not as low AND pass transistor not as big ➡ Amplifier pole at higher frequency.

(Hanumolu, JSSC 8/2008)

Now let us look at how to reduce Deterministic Jitter introduced by Power Supply noise into various PLL blocks.

We have analyzed effects of noise at various stages of the PLL to the output earlier.

The figure shows typical LDO implementation with an opamp and Pmos pass transistor driving the load.

This forms a 2 pole feedback system which effects the stability and PSRR of LDOs.

If the output pole formed by RL and CL is made dominant and the amplifier pole is kept at higher frequency, one avoids the high frequency peaking of PSRR.

Prof. Hanumolu suggests placing multiple LDOs for different blocks rather than one single for all analog blocks.

Digital PLL Motivation

25

- Multiple PLLs used in SOCs for different FUBs.
- Mostly digital process—deep submicron
- Amenable to technology porting without much redesign
- Low Power

How: Convert analog blocks like Charge Pump / Loop Filter/ VCO to digital blocks.

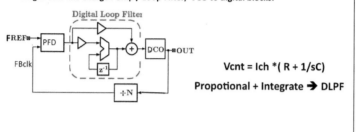

$$Vcnt = Ich *(R + 1/sC)$$

Propotional + Integrate ➔ DLPF

Most SOCs using deep submicron digital processes need multiple PLLs for different Functional blocks.

Thus, these PLLs need to have small foot print, low power and easy to port and reconfigure.

Thus, DPLL converts most analog blocks like the Charge Pump/ Loop Filter and the VCO into digital blocks.

Finally, let us briefly look at Digital PLLs techniques and trends.

The functional blocks of a DPLL are shown in this figure with details in the next few slides.

Digital PFD

26

Analog PFD: Pulse widths of UP and DN indicate amount of phase difference (Δφ ➔ΔVcnt) for APLL.

TDC type PFD: Phase difference quantized to gate delay. Linear but area/power hungry.

Digital PFD: Bang-bang type PFD where UP/DN simple indication of fast or slow (Non-linear) for DPLL.

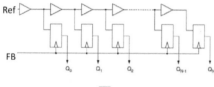

111111111111100000000   ➔ Delay= n * Tb

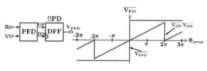

BB PFD   (Hanumolu, JSSC, 8/2011)

which reduce the quantization step to difference of delays among two types of inverters.

TDC based PFD is quite linear but its area and power are quite significant...a metric that does not go well with DPLL objectives.

A truly digital PFD, shown here, is called Bang-Bang type.

It just tells if the reference input is leading or lagging the feedback...so, the UP/DN

A Time to Digital Converter (TDC) block shown here performs similar function as the analog PFD.

The reference signal is passed thru a delay line formed by fast buffers and these delay chain is clocked into a bank of FF by the feedback clock.

The TDC provides quantized version of the phase delay in terms of number of gate delays.

Many other design techniques have come up

signal is binary. The BB PFD gain depends inversely to input phase difference.

Thus the PLL loop becomes a Non-linear feedback system. Auxiliary gain correction schemes are required for stability.

Due to small foot print and power, the BB type PFD has gained mainstay in most DPLL designs.

27

Generally for jitter <5ps, a CMOS ring oscillator with switched/selectable inverters are used as delay element.

August form Intel published this digitally controlled DCO in 2012.

Here, five tuning elements form the 5-stage ring oscillator.

It has 7b coarse control signals which can vary the frequency in 30MHz steps from 200MHz to 3.2GHz.

Then there are 10 fine control bits which change the load capacitance for each delay element with 0.025% Delta F/bit.

The coarse control bits are used for Calibration at start up while fine control bits are used for dynamic feedback control.

Digitally Controlled VCO

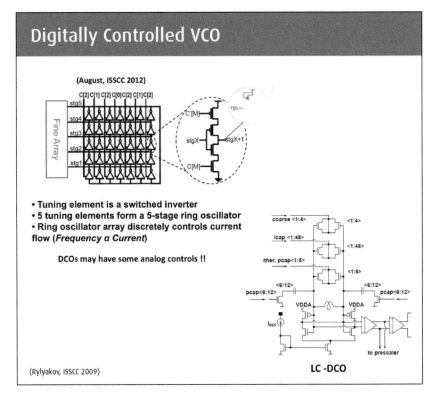

(August, ISSCC 2012)

- Tuning element is a switched inverter
- 5 tuning elements form a 5-stage ring oscillator
- Ring oscillator array discretely controls current flow (*Frequency α Current*)

DCOs may have some analog controls !!

(Rylyakov, ISSCC 2009)

LC -DCO

The DCO looks like a Memory Array in layout and takes about 80x20 um2 in 22nm CMOS technology.

When sub ps jitter is required, one has to make use of LC-VCO as shown on the right here. The capacitive DACs are used for its frequency control.

28

Hybrid DPLL Example

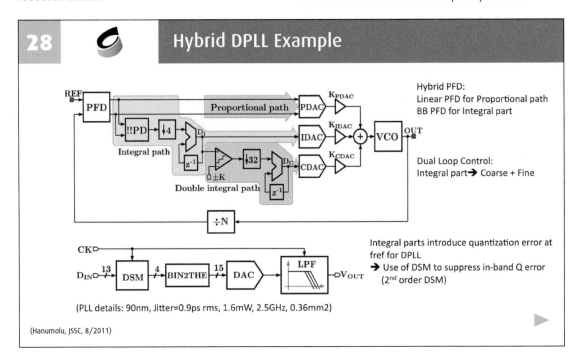

Hybrid PFD:
Linear PFD for Proportional path
BB PFD for Integral part

Dual Loop Control:
Integral part➔ Coarse + Fine

Integral parts introduce quantization error at fref for DPLL
➔ Use of DSM to suppress in-band Q error (2nd order DSM)

(PLL details: 90nm, Jitter=0.9ps rms, 1.6mW, 2.5GHz, 0.36mm2)

(Hanumolu, JSSC, 8/2011)

Hybrid DPLL Example

H ere is another example of DPLL which makes use of a traditional and BB type PFD.
So, the PLL is a hybrid of Analog and digital techniques.

The analog PFD controls the current for VCO forming the Proportional Path of VCNT.

The Integral part of VCNT makes use of BB PFD output thru an accumulator functioning as the Loop filter.

Prof. Hanumolu also included a very low frequency integrator similar to Dual Loop control system that forms the Coarse control for the VCO.

These Integral and double integral parts of the DLPF work at Feedback rate and produce high resolution control signals which would require high resolution DACs to control the VCO.

So, in most DPLLs, Delta Signal Modulators are used to interpolate these high resolution but low sample rate signals into low resolution words at close to VCO frequency.

As an example shown here for IDAC and CDACs, 2nd order DSM convert a 13b word at FB rate to 4bit at close to VCO rate.

Then, one can use a 4bit DAC to control the VCO frequency.

This PLL was implemented in 90nm for a 2.5GHz clock synthesis in 0.36mm2 with 1.6mW power only.

Summary

 29

- PLL Loop Analysis
- Analog PLL Designs
- Jitter Analysis
- Dual Loop Control
- DPLL Designs
- References

We looked at how to optimize these metrics.

We briefly looked at Dual loop control techniques where VCO gain is split between low frequency coarse control and more dynamic fine control for the lock state.

We also looked at some modern trends for Frac-N and Digital PLL techniques for SOCs.

Some of the references used for this tutorial are listed here. The last part is some back up material for more details.

I n closing, we covered PLL loop analysis.

Analog PLL designs where I showed some evolution of design techniques in DSM technologies.

Jitter, Power and Area are three critical metrics.

30 References

1. Gardner, F.M., "Charge-Pump Phase-Lock Loops", Trans. On Communications, Nov. 1980.

2. Williams et al, " An Improved CMOS Ring Oscillator PLL with less than 4ps RMS Accumulated Jitter", CICC 2004.

3. Tierno et al, "A Wide Power Supply Range, Wide Tuning Range, All Static CMOS All Digital PLL in 65nm SOI", JSSC Jan 2008

4. Hanumolu et al, "Supply Noise Mitigation Techniques in Phase-Locked Loops", JSSC, Aug 2008.

5. Wilson et al, "A CMOS Self-Calibrating Frequency Synthesizer", JSSC Oct. 2000.

6. Maxim Application Note 3359, "Clock Jitter and Phase Noise Conversion", Dec 2014

7. Young et al, " A PLL Clock Generator with 5 to 110MHz Lock Range for Microprocessors", JSSC, Dec. 1992.

8. Maneatis et al, "Self-Biased High Bandwidth Low Jitter 1 to 4096 Multiplier Clock Generator PLL", JSSC, Nov. 2003.

9. Maneatis, J.G.,"Low-Jitter Process-Independent DLL and PLL Based Self-Biased Techniques", JSSC, Nov 1996.

10. Ray et al, "A 13GHz Low Power Multi-Modulus Divider Implemented in 0.13um SiGE Technology", SiRF 2009.

11. Vaucher et al, "A Family of Low-Power Truly Modular Programmable Dividers in Standard 0.35um CMOS Technology", JSSC, July 2000.

12. Lee et al, "Oscillator Phase Noise: A Tutorial" JSSC, Mar 2000.

13. Hegazi et al, " A Filtering Techniques to Lower LC Oscillator Phase Noise", JSSC, Dec. 2001.

14.   Galton, " Delta-Sigma Fractioal –N Phase-Locked Loops" , *Phase-Locking in High-Performance Systems : From Devices to Architectures*, Wiley, 2003.

15.   Hanumolu et al, " A Calibration-Free Fractional-N Ring PLL using Hybrid Phase/Current Mde Phase Interpolation Method", JSSC, Apr. 2015.

16. Rylyakov et al, "BB DPLL at 11 and 20GHz with sub 200fs Integrated Jitter for High Speed Serial Communication Applications", ISSCC 2009.

17. August et al, " A TDC-less ADPLL with 200 to 3200MHz Range.. In 22nm CMOS", ISSCC 2012.

18. Yin et al, " A 0.7 to 3.5GHz 0.6 to 2.8mW Highly Digital PLL with BW tracking", JSSC, Aug 2011.

Back up materials

31 Dual Loop control

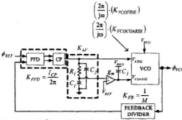

Fig 3: Improved 4th order PLL with dual control VCO

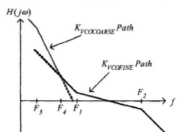

Fig. 4: Open loop transfer function of the 4th order, dual control path PLL

$$K_{LFC1}(j\omega) = K_{LF}(j\omega) \cdot \frac{1}{(j\omega\tau_1 + 1)}$$

$$K_{LFC1}(j\omega) = \frac{1}{j\omega \cdot (j\omega \cdot \tau_2 + 1) \cdot (C_1 + C_2)} \quad (4)$$

The total open loop transfer function (5) of the improved PLL can be seen in Fig. 4, where $F_3 = 1/(2\pi\tau_3)$. $K_{VCOCOARSE}$ is the voltage-to-frequency gain of the VCO's V_{COARSE} input, and $K_{VCOFINE}$ is the gain of the V_{FINE} input.

$$H(j\omega) = \left(\frac{(K_{LFC1}(j\omega) \cdot K_{GMC}(j\omega) \cdot K_{VCOCOARSE}) \cdot I_{CP}}{(j\omega) \cdot M} \right)$$

$$+ \left(\frac{(K_{LF}(j\omega) \cdot K_{VCOFINE}) \cdot I_{CP}}{(j\omega) \cdot M} \right)$$

$$H(j\omega) = \left(\frac{I_{CP} \cdot \left(\frac{g_m}{g_{out}}\right) \cdot K_{VCOCOARSE}}{(j\omega)^2 \cdot (j\omega \cdot \tau_2 + 1) \cdot (j\omega \cdot \tau_3 + 1) \cdot (C_1 + C_2) \cdot M} \right)$$

$$+ \left(\frac{(j\omega \cdot \tau_1 + 1) \cdot I_{CP} \cdot K_{VCOFINE}}{(j\omega)^2 \cdot (j\omega \cdot \tau_2 + 1) \cdot (C_1 + C_2) \cdot M} \right) \quad (5)$$

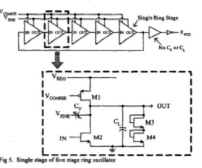

Fig 5. Single stage of five stage ring oscillator

varactor to adjust the speed of the ring. Two stacked diode connected NMOS devices (M3) and (M4) are used as a clamp

Self Calibration Technique

32

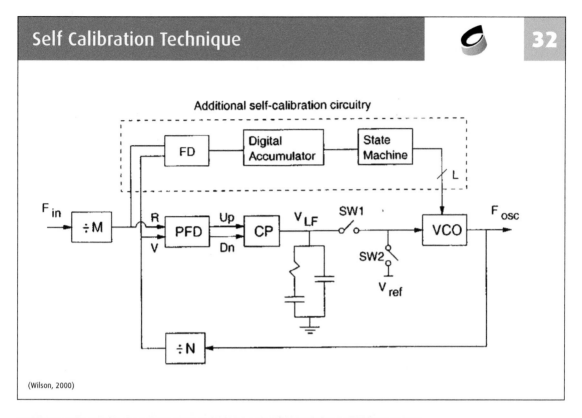

(Wilson, 2000)

Multiple LDOs (as opposed to Single)

33

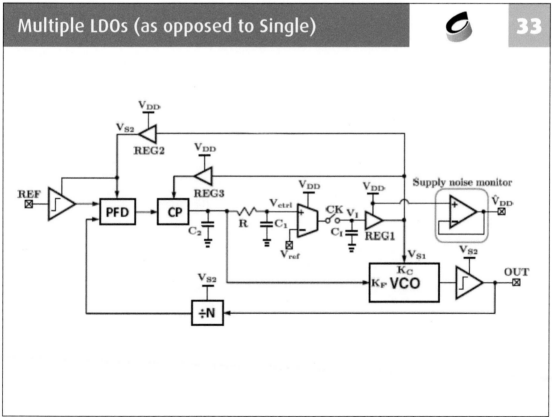

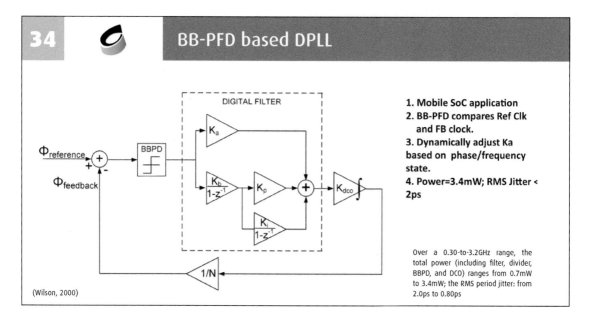

34 **BB-PFD based DPLL**

1. Mobile SoC application
2. BB-PFD compares Ref Clk and FB clock.
3. Dynamically adjust Ka based on phase/frequency state.
4. Power=3.4mW; RMS Jitter < 2ps

Over a 0.30-to-3.2GHz range, the total power (including filter, divider, BBPD, and DCO) ranges from 0.7mW to 3.4mW; the RMS period jitter: from 2.0ps to 0.80ps

(Wilson, 2000)

35 **PLL Calibration and Monitoring**

VCO Calibration-- after Power OK and before Reset release:

1. PLL is forced into Open Loop; VCO free running.
2. Ref Clk and FB Clk compared thru use of counters over a known time window and a controller sends a UP or DN signal to VCO tuning.
3. When the two counters come within specified limits, CAL engine stops and PLL is put into Closed loop to settle.

PLL Lock Monitoring:

1. FBClk continuously monitored to be within preset accuracy of Fref (same counters as in CAL)
1. If FBclk deviates out of these limits, Loss of Lock (LOL) asserted. (One may restart CAL.,

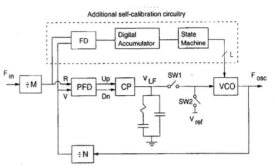

Here we have a Built In Controller for PLL Calibration and Lock Monitoring. It can also be used as BIST for PLL. Upon power UP and during Chip Reset, the PLL is forced into open loop condition. The VCO control voltage is set at Mid value thru a switch. The VCO frequency is divided down by the FB div. In the Auxiliary block for Cal/test, a FD compares input Ref CK with FB clk and the difference is accumulated over fixed time window to change the VCO frequency. Once the two counters come within specified limits, CAL engine is frozen and the PLL is put into closed loop.

Same hardware can be/is used for PLL lock monitoring. If FBclk deviates from set the limits, an LOL flag is set.

Generally, the FB clock is brought out on a GPIO to test the PLL jitter. Other analog signals such as VCNT, Iup/dn currents can also be looked at in test modes for functionality checks.

Simplified PFD/CP/LPF 36

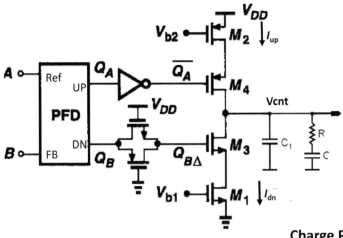

Charge Pump:

- I_{dn} must match---else static phase error!!
- Watch out for Charge Injections!
- Transmission gate equalizing delay of Inverter for UPB.

LPF:

- R and C form a zero
- C1 can be parasitic or intentional cap to filter Vcnt.
- C1 lowers phase margin slightly.

Next we move to the Charge Pump/Loop Filter. Here is a simplified schematic of PFD/CP/LPF. The Charge Pump has two equal current sources which are controlled by UP and DN signals of the PFD. We need to design Iup and Idn to be equal else one gets a static phase error in lock condition along with reference ripple on the Vcnt. The Loop filter is Type II RC filter with a ripple reducing capacitor C1. R and C form the LHP zero.

mm-Wave and Terahertz Signal Generation, Synthesis and Amplification: Reaching the Fundamental Limits

Omeed Momeni

University of California, Davis, USA

There is a growing interest in terahertz and mm-wave systems for compact, low cost and energy efficient imaging, spectroscopy and high data rate communication. Unfortunately, today's solid-state technologies including silicon and compound semiconductors have much lower performance at mm-wave frequencies compared to traditional RF bands. In order to overcome this limitation, we have introduced systematic methodologies for designing circuits and components, such as signal sources and amplifiers operating close to and beyond the conventional limits of the devices. These circuit blocks can effectively generate and combine signals from multiple devices to achieve performances orders of magnitude better than the state of the art. As an example, we show the implementation of a 482 GHz oscillator with an output power of 160 µW (-7.9 dBm) in 65 nm CMOS, a 300 GHz frequency synthesizer with 7.9% locking range in 90 nm SiGe, and a 260 GHz amplifier with a gain of 9.2 dB and saturated output power of -3.9 dBm in 65 nm CMOS.

1 **Outline**

- Motivation

- High Power Terahertz and mm-Wave Oscillators

- Wide Band Voltage Controlled Oscillators

- THz Varactor-less Scalable Standing Wave Radiator Array

- Terahertz Frequency Synthesizer

- High Gain mm-Wave and Terahertz Amplifier Design

2 **Why Higher Frequency?**

- Higher Bandwidth → Less complicated systems/Higher data rate/Higher radar resolution

Shannon-Hartley Theorem: $C = BW.log_2(1 + SNR)$

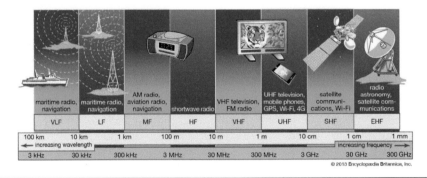

Here are a few reasons why it is necessary to move to higher frequencies for some applications. Unlike lower frequencies very wide band frequencies are available to use at mm wave and beyond. This results in less complicated systems for the same data rate or much higher data rate for the same modulation. As Shannon beautifully described, the higher the BW the higher the data rate.

Why Higher Frequency?

 3

- Less Interference → Higher allowed power → Higher SNR

- Isolation → More Frequency reuse

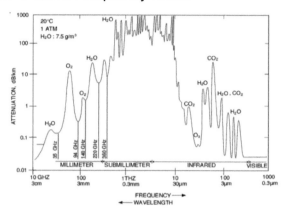

There would also be less interference as a result of high absorption of the air and water at these frequencies. Therefore higher radiated power will be allowed and that can result in higher SNR. On the other hand the better isolation enables more frequency reuse which is major challenge at lower frequencies.

Why Higher Frequency?

 4

- Unique Signatures (e.g., passes through fog, smoke, dust)

Copyright ©2006 by the University of St Andrews

The Very Large Array, an interferometric array Socorro County, New Mexico, USA

The fig on the left clearly shows the difference between mm wave and other frequency bands specially in imaging. Mm wave signal can pass through hair and can show concealed weapon underneath clothes. Moreover, about 98% of electromagnetic signals we receive from space is from 300 GHz to 3 THz. So this is an extremely important technology scientifically..

5 — Why Higher Frequency?

- Smaller Structures → System on chip

Bluetooth
Wi-Fi
GPS
UMTS
GSM

iPhone 4 GSM

Antenna size at ~2 GHz

2 mm

0.6 mm

Antenna size at ~370 GHz

The antennas and structures become smaller as the frequency increases. The figs here shows the difference in size between two antennas, one at 2 GHz and the other at 370 GHz. Because of the small structures, a complete system can be integrated on a small chip at mm wave and THz frequencies.

6 — Application of Terahertz Systems - I

- Imaging (e.g., detection of concealed weapons, cancer diagnosis, and semiconductor wafer inspection)

- Compact range radars

- High data rate communication (e.g., 100 Gbps)

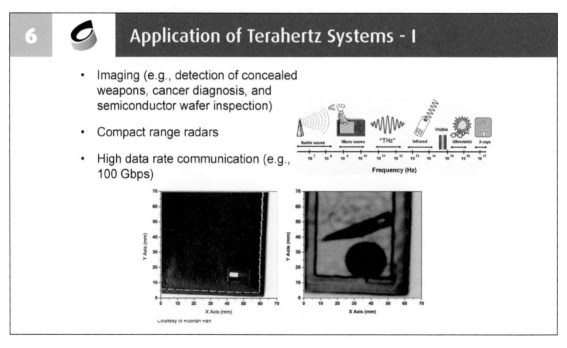

There are also lots of growing applications in this frequency range including various imaging systems and radars. The figure in right shows the detection of metallic object using 270 GHz signal.

High power sources are needed for these applications and this is the main challenge for realizing these systems.

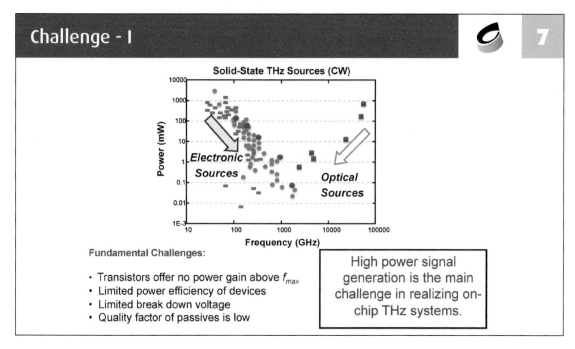

But the problem is terahertz is too low for optics and it is too high for electronics as you can see in this graph. This graph shows the maximum power generated by electronics and optics as a function of frequency. Please note that none of these sources are CMOS. So we need to use innovative techniques to boost the power in electronic side specially in CMOS, to be able to take advantage of this band efficiently.

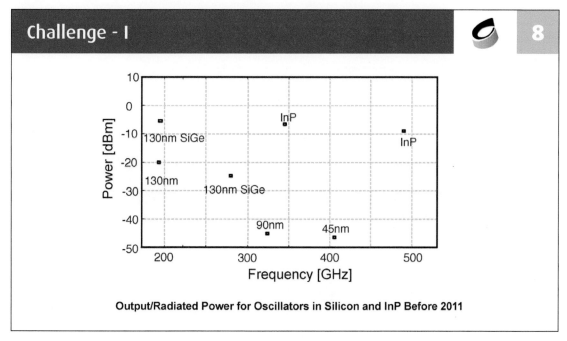

This graph shows the state of the art in mm wave and THz oscillators in 2011. There was not many sources with high power at that time. There has been a lot of research in the past few year that have improved the performances significantly.

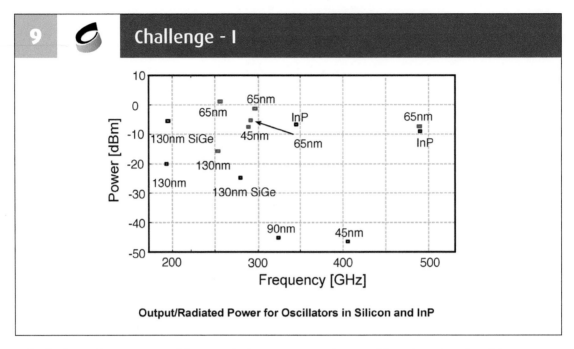

Output/Radiated Power for Oscillators in Silicon and InP

This fig shows the same state of the art in the last year. many silicon work have been introduced that generate high signal power, even more than 0 dBm. I will talk about some of these work and the reason why high power generation is now possible in silicon.

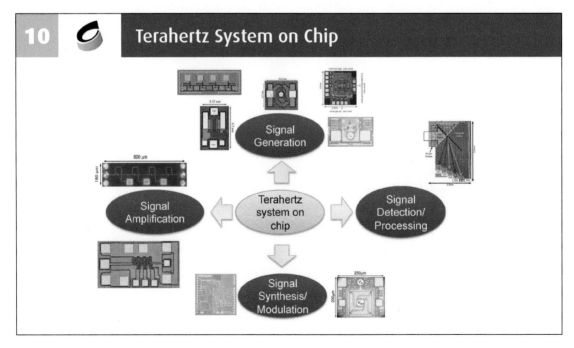

By overcoming these challenges I will present some of the essential blocks in terahertz and mm-wave frequencies that can be used in signal generation, signal detection, synthesis and amplification. This will put us one step closer to realizing a full on chip terahertz system.

Why Silicon? 11

- Enormous world wide Si Manufacturing Capacity
- Ability to integrate digital logic with RF circuits
- Scaling resulted in faster transistors

Intel Chip

t is needless to say that if we can implement any system on silicon we will certainly do that. It goes back to many advantages including the ones outlined here.

Outline 12

- Motivation

- High Power Terahertz and mm-Wave Oscillators

- Wide Band Voltage Controlled Oscillators

- THz Varactor-less Scalable Standing Wave Radiator Array

- Terahertz Frequency Synthesizer

- High Gain mm-Wave and Terahertz Amplifier Design

13   Fundamental Limits?

- Most of the fundamental oscillators have the oscillation frequency in the order of the half of the f_{max} of the transistors. Why not higher?

- What is the maximum oscillation frequency of a circuit topology, considering the quality factor of the passive components?

- For a fixed frequency, what is the topology that results in maximum output power?

As we know fmax is a fundamental figure of merit for a transistor and it is the highest frequency in which the transistor can oscillate at. In other words above this frequency the transistor is not active anymore. Given this fact, there are a few questions outlined in this slide which are important to understand. The answer to non of these questions were clear to me when I started this research.

14   Fundamental Limit: Example

- Example: IBM 130 nm CMOS process:
  - f_{max}=174 GHz (simulated)
- Regular Cross-Coupled oscillator:
  - Maximum achievable frequency (simulation): 120.7 GHz
  - This is with IDEAL inductors!

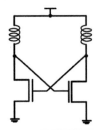

To make this questions more clear we performed a very simple simulation. An IBM 130 nm CMOS transistor has a simulated fmax of 174 GHz. We used this transistor in a cross coupled oscillator with ideal passive components. We reduced the size of the inductors to increase the oscillation frequency thinking that the oscillation frequency should reach 174 GHz. However the max frequency was 120 GHz. How is that the case? Why cant we reach 170 GHz in a cross couple oscillator?

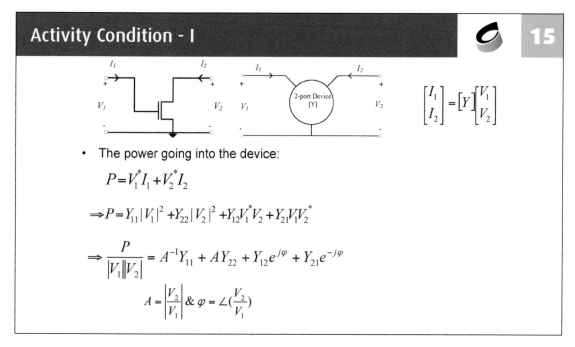

Activity Condition - I   15

- The power going into the device:

$$P = V_1^* I_1 + V_2^* I_2$$

$$\Rightarrow P = Y_{11}|V_1|^2 + Y_{22}|V_2|^2 + Y_{12}V_1^*V_2 + Y_{21}V_1V_2^*$$

$$\Rightarrow \frac{P}{|V_1||V_2|} = A^{-1}Y_{11} + AY_{22} + Y_{12}e^{j\varphi} + Y_{21}e^{-j\varphi}$$

$$A = \left|\frac{V_2}{V_1}\right| \ \& \ \varphi = \angle\left(\frac{V_2}{V_1}\right)$$

To find the answer to this question we looked into the basic power flow in a single transistor. We found the power flowing out of any transistor as a function of the Y parameters of the transistor and the voltage gain and phase shift of the device, A and phi.

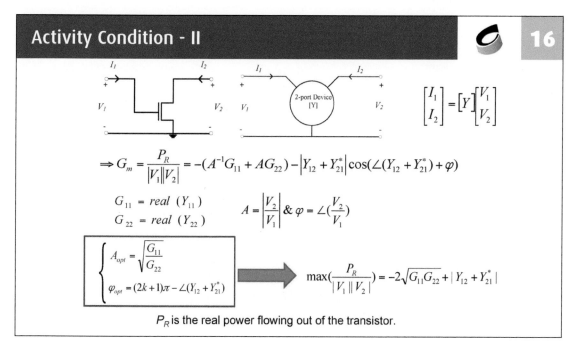

Activity Condition - II   16

$$\Rightarrow G_m = \frac{P_R}{|V_1||V_2|} = -(A^{-1}G_{11} + AG_{22}) - |Y_{12} + Y_{21}^*|\cos(\angle(Y_{12} + Y_{21}^*) + \varphi)$$

$$G_{11} = real\ (Y_{11})$$
$$G_{22} = real\ (Y_{22})$$

$$A = \left|\frac{V_2}{V_1}\right| \ \& \ \varphi = \angle\left(\frac{V_2}{V_1}\right)$$

$$\begin{cases} A_{opt} = \sqrt{\dfrac{G_{11}}{G_{22}}} \\ \varphi_{opt} = (2k+1)\pi - \angle(Y_{12} + Y_{21}^*) \end{cases}$$

$$\Longrightarrow \quad \max\left(\frac{P_R}{|V_1||V_2|}\right) = -2\sqrt{G_{11}G_{22}} + |Y_{12} + Y_{21}^*|$$

P_R is the real power flowing out of the transistor.

We can then find the normalized real power as a function of the same parameters. We can then find the optimum A and phi that results in maximize real power flowing out of the transistor.

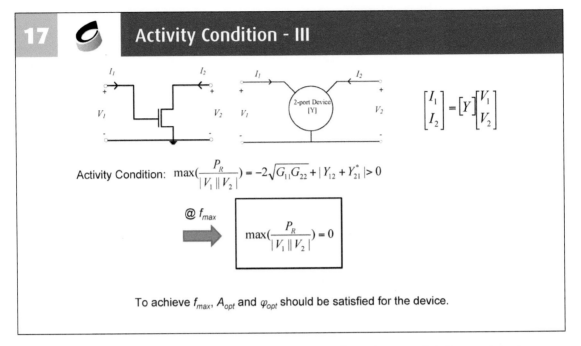

17 **Activity Condition - III**

Activity Condition: $\max(\dfrac{P_R}{|V_1||V_2|}) = -2\sqrt{G_{11}G_{22}} + |Y_{12} + Y_{21}^*| > 0$

$$\begin{bmatrix} I_1 \\ I_2 \end{bmatrix} = [Y]\begin{bmatrix} V_1 \\ V_2 \end{bmatrix}$$

@ f_{max} $\longrightarrow$ $\max(\dfrac{P_R}{|V_1||V_2|}) = 0$

To achieve f_{max}, A_{opt} and φ_{opt} should be satisfied for the device.

The transistor is active only if this maximum power is positive. This is the exact definition we have for power when we reach the fmax of the transistor. This means that if we like to reach fmax in in oscillators we need to satisfy the optimum voltage gain and phase for the transistors. These conditions were clearly not satisfied in a cross coupled oscillator for the 130 nm CMOS transistors.

18 **Example : Ring Oscillator**

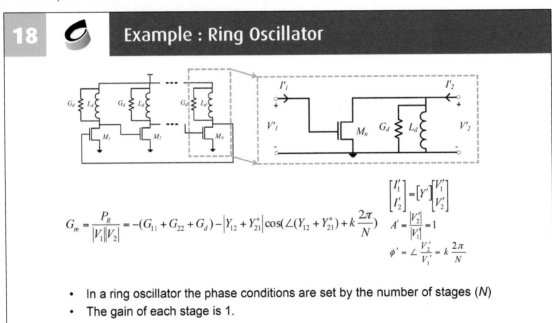

$$\begin{bmatrix} I'_1 \\ I'_2 \end{bmatrix} = [Y']\begin{bmatrix} V'_1 \\ V'_2 \end{bmatrix}$$

$$G_m = \frac{P_R}{|V_1||V_2|} = -(G_{11} + G_{22} + G_d) - |Y_{12} + Y_{21}^*|\cos(\angle(Y_{12} + Y_{21}^*)) + k\frac{2\pi}{N}$$

$$A' = \frac{|V'_2|}{|V'_1|} = 1$$

$$\phi' = \angle\frac{V'_2}{V'_1} = k\frac{2\pi}{N}$$

- In a ring oscillator the phase conditions are set by the number of stages (N)
- The gain of each stage is 1.

To verify this theory we chose ring oscillator because first it is easy to manipulate the phase conditions for each stage by changing the number of stages and second the cross coupled oscillator is a 2 stage ring and its properties can be analyzed by this.

Optimum Conditions

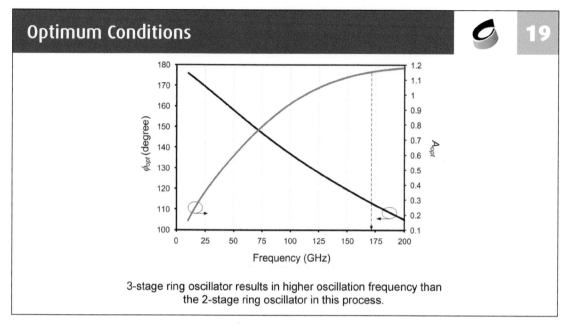

3-stage ring oscillator results in higher oscillation frequency than
the 2-stage ring oscillator in this process.

We can plot the optimum conditions as a function of frequency for any given transistor. Here the optimum conditions are plotted for the same 130 nm CMOS transistor. In order to reach fmax of 174 GHz we need to satisfy the conditions at that frequency. It is easy to see that 3 stage ring oscillator satisfies these conditions and hence can reach frequencies very close to fmax. These conditions can be different in different processes and for different sizes as the transistor properties change.

Design Example: mm-Wave Oscillators

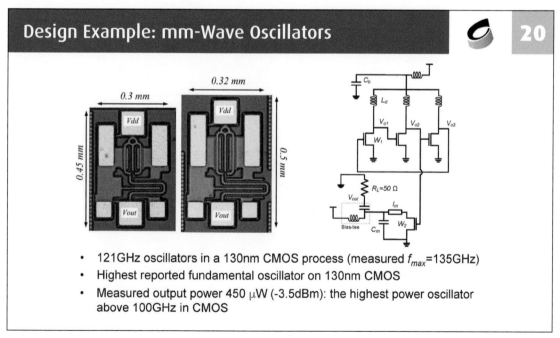

- 121GHz oscillators in a 130nm CMOS process (measured f_{max}=135GHz)
- Highest reported fundamental oscillator on 130nm CMOS
- Measured output power 450 μW (-3.5dBm): the highest power oscillator above 100GHz in CMOS

To experimentally verify the design a 121 GHz 3-stage oscillator was fabricated. The oscillation frequency was very close to the measured fmax of 135 GHz.

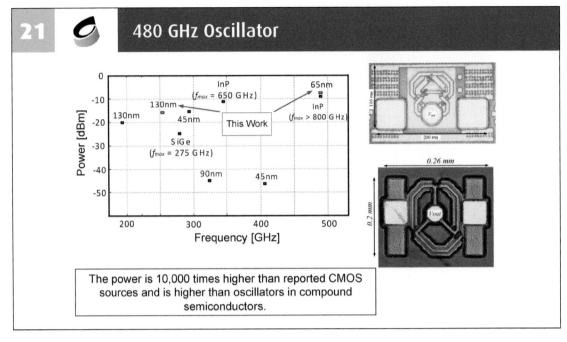

21   480 GHz Oscillator

The power is 10,000 times higher than reported CMOS sources and is higher than oscillators in compound semiconductors.

Using the same design methodology a 480 GHz harmonic oscillator was also fabricated in a 65 nm CMOS process with -7.9 dBm output power. The generated power is still the highest among oscillators at the same frequency range.

22   Outline

- Motivation

- High Power Terahertz and mm-Wave Oscillators

- Wide Band Voltage Controlled Oscillators

- THz Varactor-less Scalable Standing Wave Radiator Array

- Terahertz Frequency Synthesizer

- High Gain mm-Wave and Terahertz Amplifier Design

Motivation 23

- Applications for wideband signal generation at mm-wave and THz frequencies include

  - Spectroscopy
    - Wideband signals for signature detection of materials
  - High data-rate communication
    - Data-rate is a direct function of bandwidth
  - High resolution radar
    - Range resolution is a function of bandwidth
    - Azimuth resolution is a function of frequency

- Demand wideband and powerful Signal sources

Mm-wave and THZ application such as spectroscopy, high-date-rate communication and high-resolution radar require wideband signal sources. For example, signature detection of materials needs wide bandwidth. Same applies to range resolution in radars and communication.

mm-wave and THz Signal Sources 24

- Frequency Multiplier/amplifier chain

  - Wider bandwidth
  - High conversion loss
  - More power consumption
  - Larger chip area

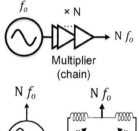

✓ **Harmonic VCOs**
  - Lower tuning range
  - Less power consumption
  - Smaller chip area

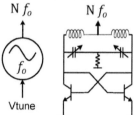

The proposed circuit has significantly improved the tuning range

There are two main approaches to generate signal at frequencies above 100GHz. One, is to use multiplier/amplifier chain and Two, is to use harmonic VCOs. Although multiplier chain have higher bandwidth, Its drawbacks such as high conversion-loss, high power consumption and Large chip area makes tuning range of harmonic VCOs a hot topic for researchers. We choose to boost the tuning range of the harmonic oscillators to achieve low power consumption and small chip area at the same time.

25 Challenges of Tuning Range

- Low quality factor of varactors
  - Impedes Oscillation start up
  - Lower oscillation swing
  - Lower harmonic power

- Parasitic capacitance of transistors
  - Comparable to varactors
  - Limits the tuning range

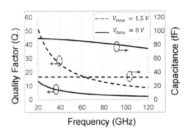

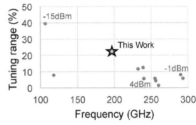

There are two major challenges for Frequency-tuning at frequencies above 100GHz. One, is the low quality fctor of the varactors. The left figure shows the quality factor of a typical varactor in 0.13um CMOS process where its quality factor varies between 10 to 3 at 100GHz. This lowers the oscillation swing and harmonic power.

Moreover, large parasitic capacitors of transistors is comparable to varactors and therefore, tuning range is limited. The figure on the right marks the previous reported works where the tuning ranges are less than 13% at frequencies above 120GHz.

26 Proposed Approach

- A novel **Active Mode Switching (AMS)** block is used to couple two identical VCOs
  - Each mode has separate center frequency
  - Tuning in each mode
  - Overlap between the two modes

- Low-loss mode-switching
  - Strong oscillation for high output power

- Low-Cap. mode-switching
  - High-frequency of operation
  - Wide tuning range

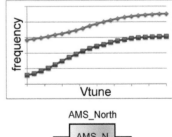

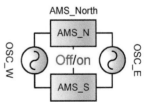

This work proposes a novel Active-Mode-Switching block, or in short AMS which assists with mode-switching between two core oscillators. The circuit operates in two distinct modes. Their bandwidth overlap for a continuous wider tuning-range. The AMS blocks do low-loss switching. This helps with strong oscillation and high output power. AMS blocks also have low parasitic capacitors which assists with high-frequency of oscillation and wide tuning range.

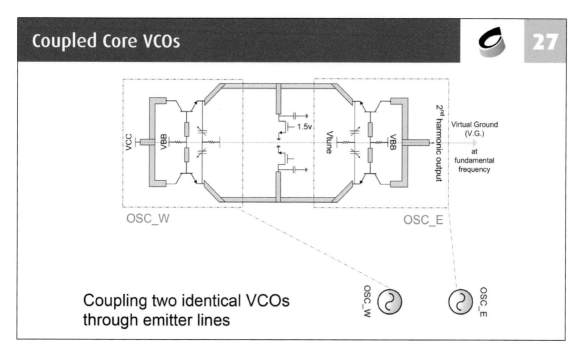

Coupled Core VCOs   27

Coupling two identical VCOs
through emitter lines

Here are two of the core VCOs. They have a differential Colpitts structure. SECOND harmonic is extracted from the collectors. Varactors are in emitter; therefore, they are not in harmonic extraction path. It turns out HBT transistors in such a structure helps a lot with the tuning range. There are two of these core oscillators; Oscillator_WEST and Oscillator_EAST. They are coupled with their emitter lines.

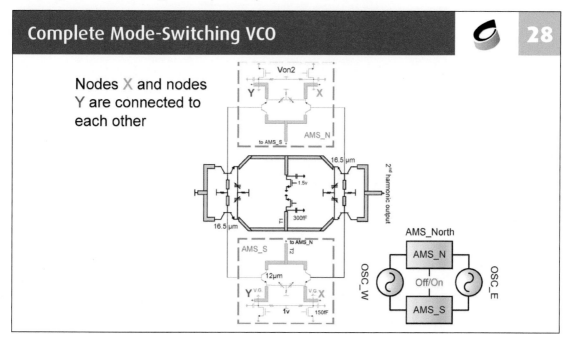

Complete Mode-Switching VCO   28

Nodes X and nodes
Y are connected to
each other

Active Mode Switching blocks called as AMS lie between the two core oscillators. They are connected directly to the tank of the core oscillators at the bases of the transistors.

These AMS blocks are varactor degenerated and have a Colpitts structure; Very similar to the core oscillator. By turning them on or off, mode of operation is determined.

29   Even Mode: In-phase Operation

- AMS blocks are turned-off
- T1 forces oscillation to be in-phase
- T1 helps with stronger oscillation

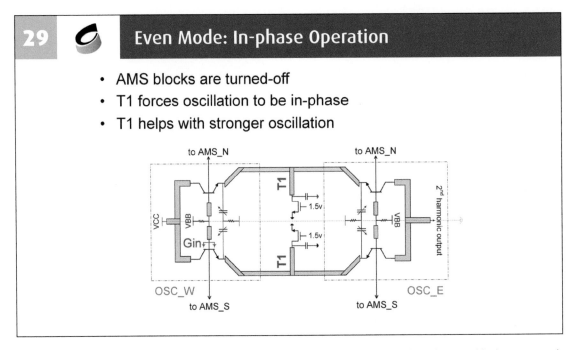

This figure shows the two core VCOs which share the line T1 at their emitters. When the AMS blocks are turned-off, In the case of in-phase oscillation, this Line helps the transistors with better loss-cancelling at the tanks. Therefore, oscillation is forced to be in-phase.

30   Even Mode: AMS Equivalent Circuit

- In-phase oscillation
  - Varactors are in series with 4kΩ resistors
  - Varactors loadings on the tanks are reduced

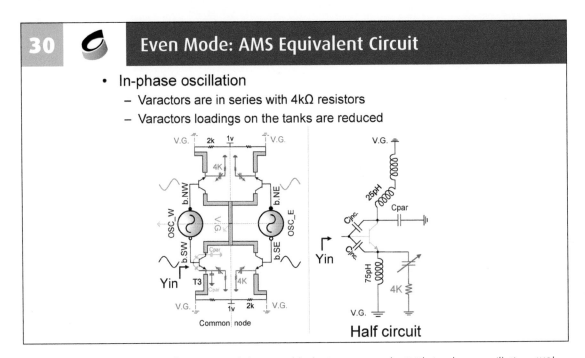

Half circuit

This figure shows the equivalent circuit of the AMS blocks in even mode. With in-phase oscillation 4KOhm resistors are in series with the varactors. This reduces the lossy and capacitive effect of the varactors on the tanks when looking into the bases of the AMS blocks.

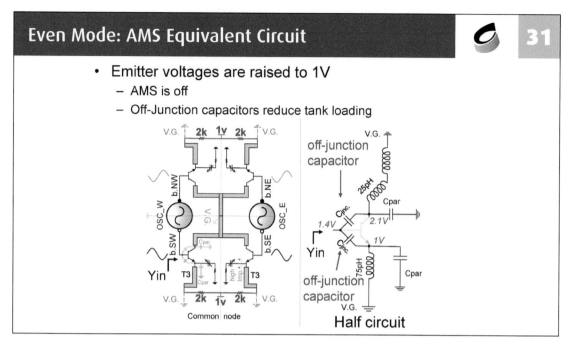

In this mode, AMS blocks are turned-ff. The emitter voltages are pulled up to 1V using the 2KOhm resistors in red. They are located at the virtual ground nodes of the circuit. With the DC-biasing voltages shown in the right, the base-collector and base-emitter voltages are small off-junction capacitors. This reduces capacitive loading of each of the AMS transistors on the tanks.

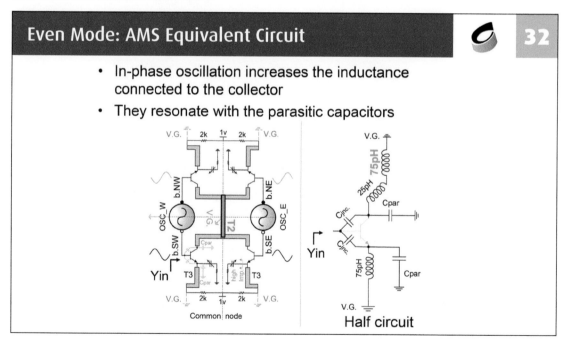

Furthermore, with In-Phase oscillation, the two transistors in each of AMS block share the green line T2. This increases the inductance connected to the collector of each of the AMS transistors. These large inductors resonate with the small parasitic capacitors.

33 Even Mode: AMS Loading on Tank

- Large inductors at collector and emitter
- Small off Junction Capacitors
- Large resistors in series with the varactors

 They lower the loss and capacitance loading of the AMS blocks
Tuning range, output power and frequency are kept high

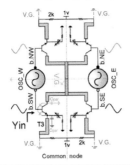
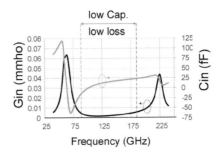

Adding up all these effects, AMS blocks load the tanks with low-loss and low-Capacitance admittance. This is shown on the right figure. Less than 10fF capacitance and less than 4mmho conductance. Therefore, tuning range and output power are kept high simultaneously.

34 Odd Mode: AMS Equivalent Circuit

- **AMS is on**
- **With out-of-phase oscillation, varactors are seen from the tanks**

  - AMS transistors are now Colpitts embedded
  - AMS cancels loss at tank
  - varactors contribute to frequency tuning

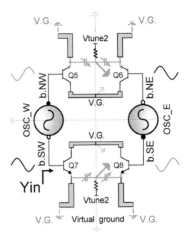

In odd mode, AMS Transistors are turned-on. In the case of out-of-phase oscillation, varactors have virtual ground between them. This virtual ground also helps with loss-cancelling at the tanks by embedding the transistors in a Colpitts structure. This forces out-of-phase oscillation for stronger oscillation. Varactors also contribute to tuning range in this mode.

Odd Mode: AMS Equivalent Circuit

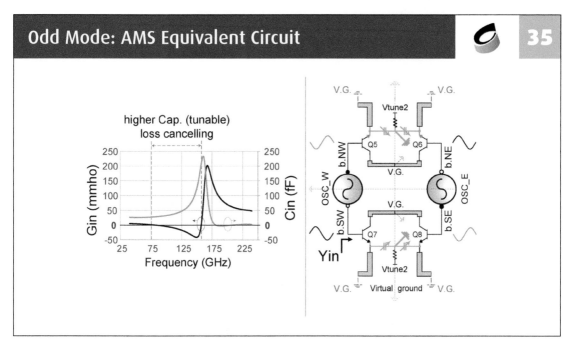

Here is the simulated input admittance, looking into the base of one of the AMS transistors. Negative conductance helps with a stronger oscillation and higher capacitance lowers the frequency of operation.

Chip Photo

- Designed and implemented in IBM 0.13μm BiCMOS process
- All the lines are microstrip using top two metal layers and were fully simulated

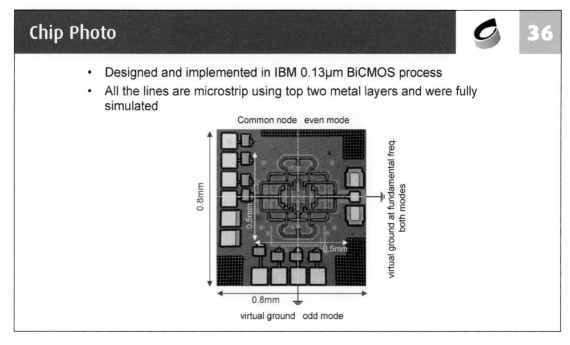

Chip photo is shown here. Layout is symmetric.The horizontal line shows the always-Virtual Ground nodes in both of the modes. The dashed vertical line shows the virtual ground noes only in odd mode.

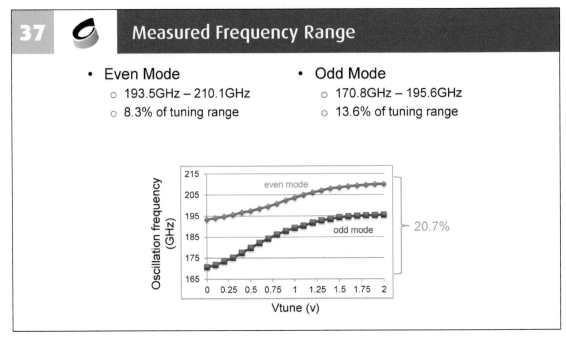

This figure shows the measured frequencies at each mode. Even mode has excellent tuning range due to the low-capacitance behavior of the AMS blocks in this mode. Odd mode has higher tuning range because of the aid from the varactors in the AMS blocks.

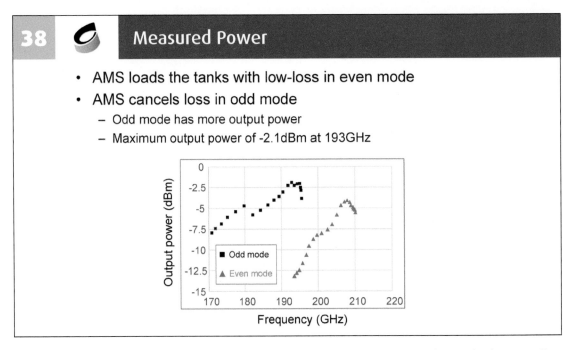

This is the measured output power at each mode. Odd mode has more power due to the loss-cancelling behavior of the AMS blocks in this mode

Nevertheless, Even mode also has excellent output power due to the low-loss behavior of the AMS blocks.

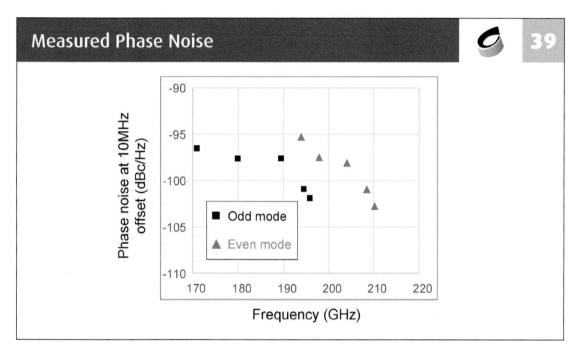

H ere is the measured phase noise at 10MHZ offset for different frequencies in each mode.

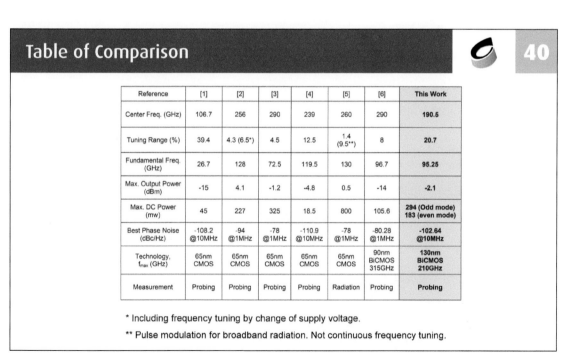

Reference	[1]	[2]	[3]	[4]	[5]	[6]	**This Work**
Center Freq. (GHz)	106.7	256	290	239	260	290	**190.5**
Tuning Range (%)	39.4	4.3 (6.5*)	4.5	12.5	1.4 (9.5**)	8	**20.7**
Fundamental Freq. (GHz)	26.7	128	72.5	119.5	130	96.7	**95.25**
Max. Output Power (dBm)	-15	4.1	-1.2	-4.8	0.5	-14	**-2.1**
Max. DC Power (mw)	45	227	325	18.5	800	105.6	**294 (Odd mode) 183 (even mode)**
Best Phase Noise (dBc/Hz)	-108.2 @10MHz	-94 @1MHz	-78 @1MHz	-110.9 @10MHz	-78 @1MHz	-80.28 @1MHz	**-102.64 @10MHz**
Technology, f_{max} (GHz)	65nm CMOS	65nm CMOS	65nm CMOS	65nm CMOS	65nm CMOS	90nm BiCMOS 315GHz	**130nm BiCMOS 210GHz**
Measurement	Probing	Probing	Probing	Probing	Radiation	Probing	**Probing**

* Including frequency tuning by change of supply voltage.

** Pulse modulation for broadband radiation. Not continuous frequency tuning.

T his table compares our work with the state-of-the-art. Our tuning range of 20.7% is significantly higher than the best reported of 12.5% for frequencies above 120GHz. which is significantly higher among all reported works above 120GHz which used to be 12.5%. The output power of -2.1 dBm is also competitive with other works.

41 **Outline**

- Motivation

- High Power Terahertz and mm-Wave Oscillators

- Wide Band Voltage Controlled Oscillators

- THz Varactor-less Scalable Standing Wave Radiator Array

- Terahertz Frequency Synthesizer

- High Gain mm-Wave and Terahertz Amplifier Design

42 **Challenges**

❑ f_{max} of transistors

    ❑ Limited power generation

    ❑ Harmonic osc. to go beyond f_{max}

    ❑ Must employ *arrays* of radiators

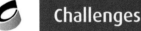

f_{max} of the 130nm HBT transistor

$f_{max} \approx 215\,GHz$

❑ Losses & parasitics of coupling networks

    ❑ Degrades power generation and operating freq.

❑ Q of varactors

    ❑ Limits freq. tuning range & output power

As we have to use harmonic oscillators to go beyond fmax. To achieve a reasonable power level, we will need to implement coherent arrays. The coupling networks required to sync the oscillators in an array will also add losses and parasitics that limit frequency tuning and power. The other issue is the poor quality factor of varactors that significantly limits the frequency tuning and their large losses would further degrade the generated power.

Proposed Approach

43

- ❑ Employ <u>standing wave distributed oscillator</u> as array's building block

- ❑ Scalable uni-structure distributed array

  - ❑ Array extension by replication of unit cell

  - ❑ Inherently in phase

  - ❑ Structural coherency

- ❑ Tuning range extended

  - ❑ Avoiding varactors

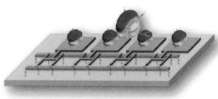

Illustration of the proposed architecture

I n this work, we have proposed a new scalable array architecture based on standing waves. This allows us to implement a uni-structure array that can be scaled to create coherent radiation. We were also able to extend the tuning range by an alternative method while avoiding varactors.

Standing Wave Oscillator (SWO)

44

- ❑ Transistors provide negative g_m

  - ▪ Compensate for losses

  - ▪ Sustain oscillation

- ❑ Terminations provide complete reflection of travelling waves

  - ▪ Back and forth waves form SW

  - ▪ Position dependent amplitude

  - ▪ Nodes/Anti-nodes

  - ▪ In-phase/out-of-phase operation

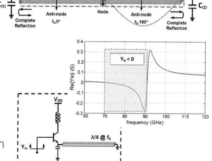

The top figure shows the structure of a simple standing wave oscillator. The capacitively degenerated transistor, shown in the bottom figure, provides narrowband negative transconductance at its base terminal that compensates for the losses in the circuit and sustains the oscillation. This creates a travelling wave that propagates through the transmission line and is reflected back and forth by the terminations at the ends. The superposition of these forward and reflected waves form a standing wave on the line. The amplitude of the signal on the line is position dependent. At the nodes with zero amplitude, the forward and reflected waves cancel each other. These waves add up constructively at the anti-nodes and results in maximum oscillation amplitude at the base of the transistors. The distance between adjacent nodes are lambda over two and all the points within that section are in phase with each other and out of phase with the section next to that.

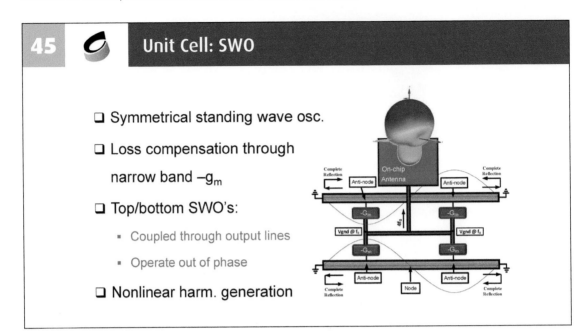

The figure on the right shows the unit cell of the proposed array. It is a symmetrical standing wave oscillator that consists of two oscillators facing each other and coupled together through the output lines. These two oscillators are forced to operate out of phase to create a virtual ground in the middle node and provide the necessary negative gm for oscillation. In addition to compensating for loss, the nonlinear narrow band negative gm generates the 4th harmonic which is radiated by an on chip antenna.

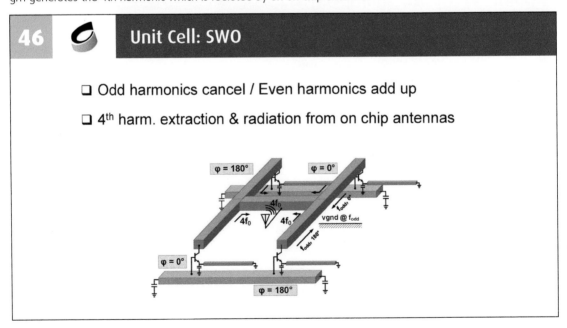

The physical implementation of this unit cell is shown here. Each transistor will be out of phase with its neighboring transistors. As a result, the odd harmonics, including the large unwanted fundamental signal, will be canceled in the middle node. The generated even harmonics, on the other hand, will be in phase and add up at the output. The desired fourth harmonics will be combined, extracted and fed to the on chip antenna for radiation. All the other even harmonics are heavily suppressed by the output lines and the antenna design and matching network.

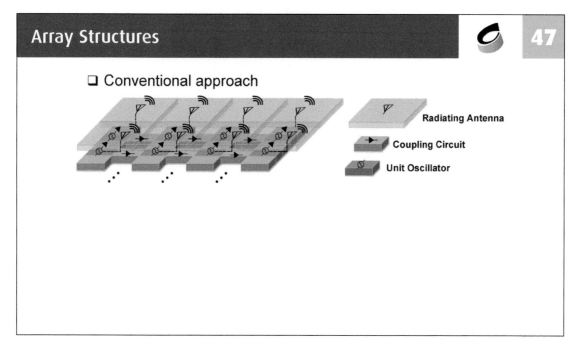

In a conventional array, to create coherent radiation, the independent oscillators in blue tiles, are coupled together with coupling networks, shown by the green tiles. These coupling networks add extra parasitics and losses to the circuit which degrades the output power and reduces the frequency of operation.

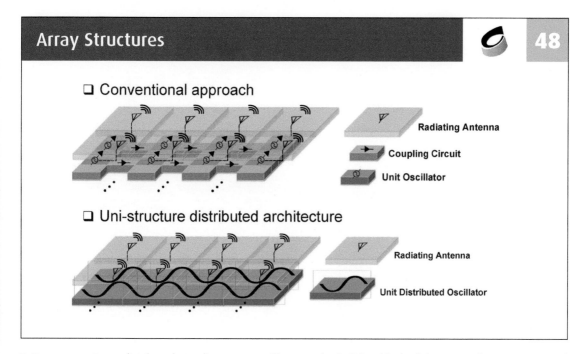

However, using a distributed standing wave oscillator as the building block of the array allows us to extend the array by simply replicating it. This would create a continuous long standing wave on the structure that generates constructive even harmonics. This way we would be able to avoid the lossy and parasitic coupling networks and scale the array as large as we want.

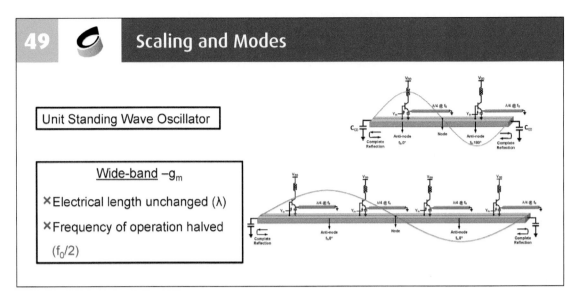

Nevertheless, simply extending the standing wave length might result in undesired modes. Consider the unit standing wave oscillator on top. In the bottom figure, we have simply extended the oscillator length by replicating the unit cell. If, however, the oscillation can be sustained at half the original frequency, f0 over two, it would simply move to that operation mode. In this case, the electrical length of the line would remain at one wavelength but the frequency would be divided by two. That requires the transistor to be able to provide negative gm at f0 over two. If the negative gm is wideband enough, like the conventional cross coupled, there would be a change in operation mode which is not desired.

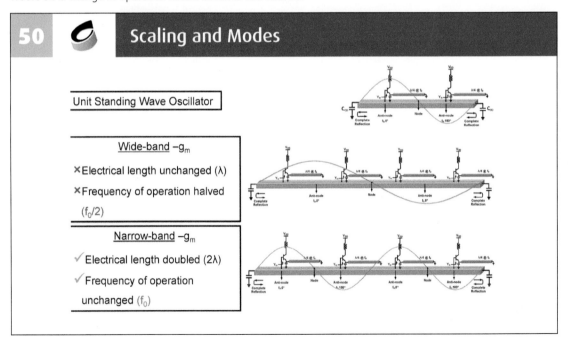

On the other hand, if the negative gm is narrow band, as is in our case, the oscillation would not be able to start up and continue at f0 over two. So the circuit is forced to remain in the original oscillation frequency but double its electrical length to two lambda as shown in the bottom figure. So, a narrow band negative gm is necessary to prevent undesired modes when scaling the oscillator into an array.

1X4 SW Radiator Array   51

❑ The proposed 1X4 scalable standing wave array

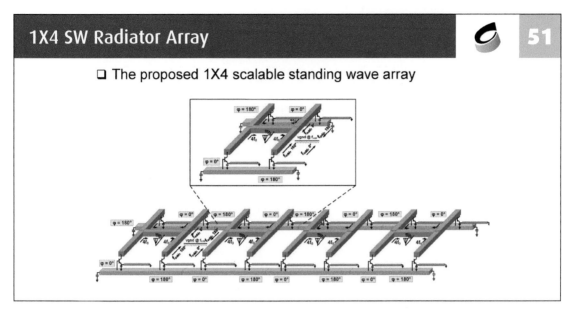

Based on the mentioned approach, we designed a four element array by extending and replicating the unit standing wave oscillator.

Two separate continuous standing waves will form on the top and bottom lines with four lambda lengths. These waves will oscillate out of phase with each other and produce in phase fourth harmonics. The continuous standing waves on each line ensure that the power fed to all four antennas are in phase.

This structure can be scaled to more radiation elements by more replicas of the unit cell.

Radiation Structure Simulation   52

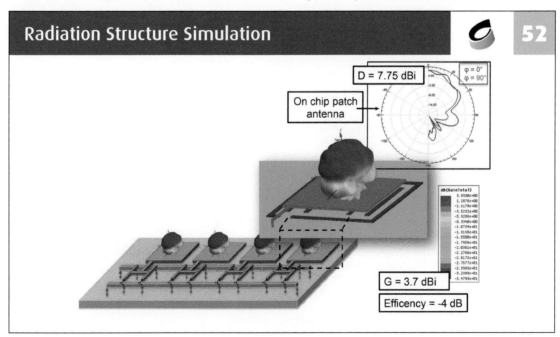

The bottom figure shows the floorplan of the structure. We used on chip patch antennas to radiate the extracted fourth harmonic power due to their single feed architecture and their good efficiency and directivity. It also allows us to radiate from front side of the chip and avoid post processing of the substrate or having to use a silicon lens. The antennas are characterized in a 3D EM simulator.

The simulated single antenna has 7.7 dBi maximum directivity and about 40% efficiency.

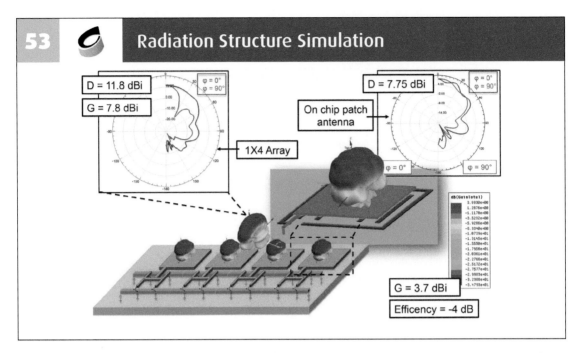

53 Radiation Structure Simulation

The simulated radiation pattern of the array in the far field is shown here. It has a maximum directivity of 11.8 dBi.

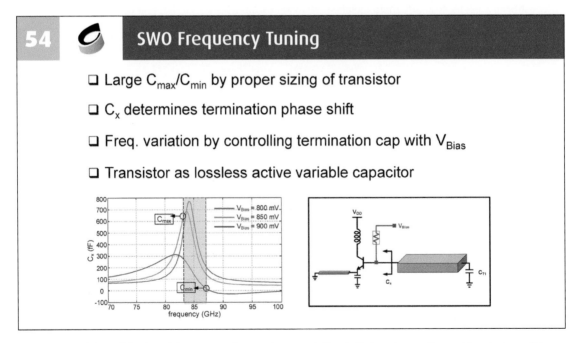

54 SWO Frequency Tuning

❏ Large C_{max}/C_{min} by proper sizing of transistor

❏ C_x determines termination phase shift

❏ Freq. variation by controlling termination cap with V_{Bias}

❏ Transistor as lossless active variable capacitor

By proper sizing of the transistor, we can have a large variation in its input capacitance. So what we did was to use the bias voltage of the transistor as a control voltage to change Cx. This will change the termination capacitance and its phase shift and therefore the frequency of the oscillator. So the frequency tuning is done by changing the termination capacitance and the transistor will provide an active variable capacitor at the termination without loss. However, we need to examine the effect of changing the bias voltage of the transistors on the output power.

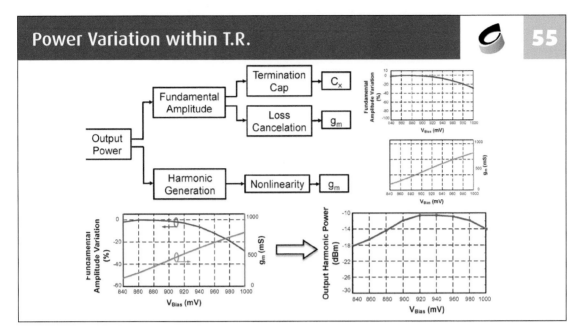

As we increase the transistor bias, the fundamental signal amplitude will drop but the transistor harmonic generation will go up. Theses two affect the output power in opposite directions. The red curve shows the simulated output harmonic power versus bias voltage and it shows that we can have reasonable power variation across our entire bias range. So the trade off between the termination capacitance and harmonic generation in a standing wave oscillator enables us to extend the tuning range with reasonable variations in output power.

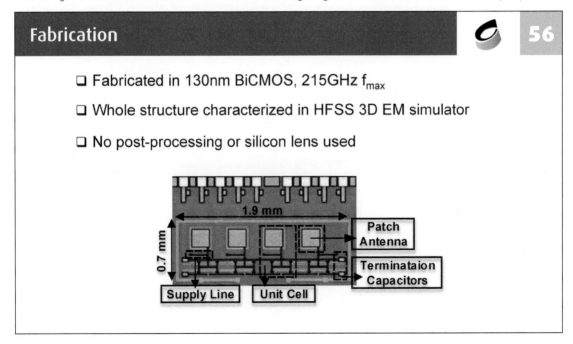

We implemented the array in a 130nm BiCMOS process with 215 GHz fmax. The antennas and passive networks were characterized in HFSS 3D EM simulator. No post processing was done and the radiation was measured without an additional silicon lens.

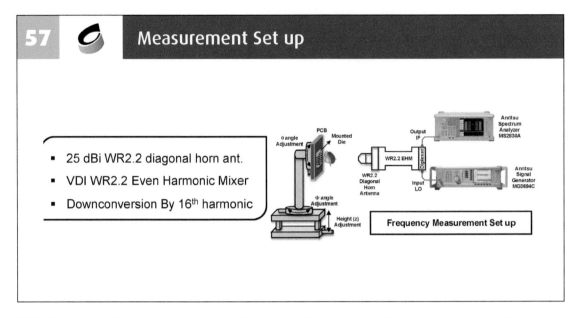

57 **Measurement Set up**

- 25 dBi WR2.2 diagonal horn ant.
- VDI WR2.2 Even Harmonic Mixer
- Downconversion By 16th harmonic

Frequency Measurement Set up

This figure shows the measurement set up that we used for frequency characterization and radiation pattern measurement. The chip was mounted on a low cost FR4 board and DC voltages were provided by simple bond wires. The board was attached to a precision mechanical positioner to locate the beam and measure the radiation pattern of the chip. At the receiver side, a diagonal horn antenna captures the beam and feeds it to a WR2.2 even harmonic mixer. The mixer downconverts the THz signal to an IF frequency that is observable on the spectrum analyzer.

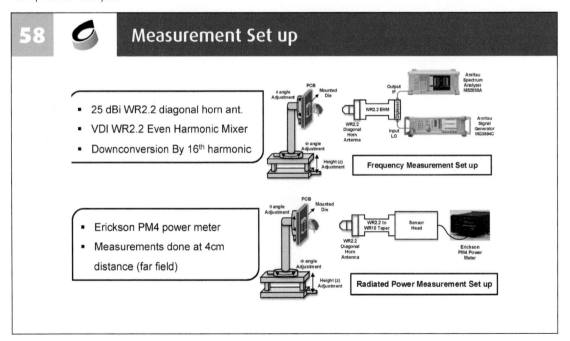

58 **Measurement Set up**

- 25 dBi WR2.2 diagonal horn ant.
- VDI WR2.2 Even Harmonic Mixer
- Downconversion By 16th harmonic

Frequency Measurement Set up

- Erickson PM4 power meter
- Measurements done at 4cm distance (far field)

Radiated Power Measurement Set up

To measure the radiated power from the array, an erickson PM4 power meter was used at 4cm distance to ensure far field.

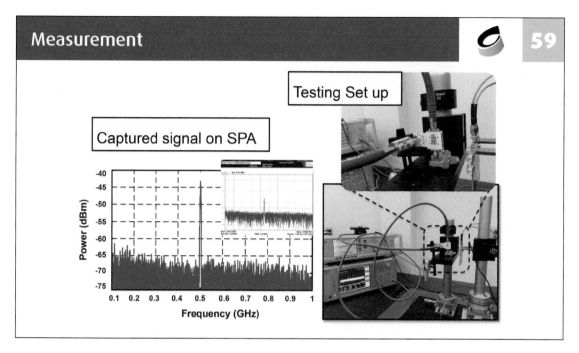

Here, you can see our actual measurement set up on the right and the spectrum of the detected radiation signal observed on the spectrum analyzer after downconversion by 16th harmonic of the LO on the left.

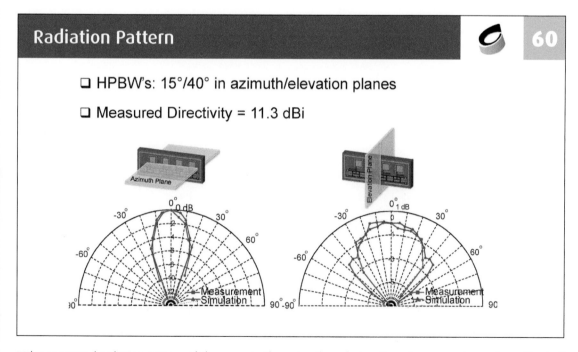

The measured radiation pattern of the array in the azimuth and elevation planes are shown here. The half power beam widths are 15 and 40 degrees respectively which result in a measured maximum directivity of 11.3 dBi.

61 Frequency and Power

❑ 5.5 dB power variation from 334GHz to 352GHz

❑ -10.5 dBm total radiated power

❑ 425mW from 1.8V supply

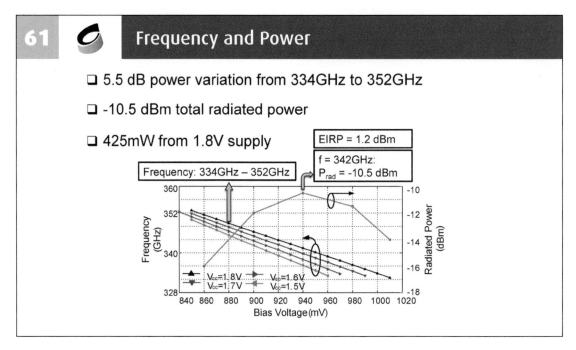

The frequency tuning and radiated power measurements are shown here. The measured radiated power showed less than 6 dB variation across the tuning range from 334 to 352 GHz. The total radiated power from the 4 elements was measured to be -10.5 dBm at 342 GHz. The chip consumes a total of 425mW from a 1.8V supply voltage.

62 Phase Noise

❑ -98.2 dBc/Hz phase noise at 10MHz offset at center freq.
  ▪ Including the noise added by the receiver

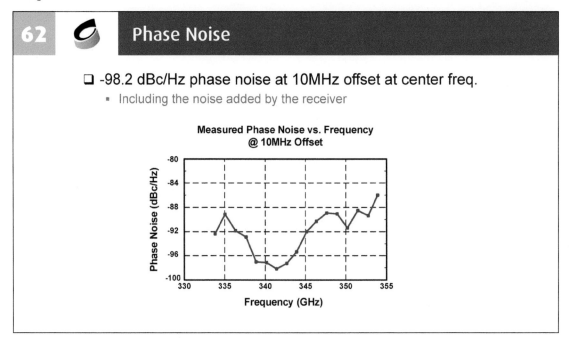

Here is the measured phase noise of the radiated signal across the tuning range at 10 MHz offset frequency. The phase noise at center frequency is equal to -98.2 dBc/Hz. This value includes the extra noise added by the harmonic mixer and the IF amplifier used for this measurement.

Performance Summary

 63

	Freq.	Technology	f_{max} (GHz)	Size	Si Lens	Tuning Range (%)	P_{rad} (dBm)	P_{rad}/Cell (dBm)	EIRP (dBm)	DC Power (W)
This work	342	130nm SiGe	215	1x4	No	5.9	-10.5	-16.5	1.2	0.425
[1]	320	130nm SiGe	280	4x4	No	N/A	0.9	-11.1	13.9	0.61
					Yes		5.2	-6.8	22.5	
[2]	338	65nm CMOS	250	4x4	No	2.1	-0.9	-13	17.1	1.54
[3]	530	130nm SiGe	500	4x4	Yes	3.2	0	-12	25	2.54
[4]	280	CMOS 45nm SOI	N/A	4x4	No	3.2	-7.2	-19.2	9.4	0.81

❑ To our knowledge, this is the largest tuning range reported for a radiating array above 300GHz

ere is a performance summary of the chip. We were able to extend the tuning range to 5.9%, which to our knowledge is the largest value reported for an array above 300 GHz.

Outline

 64

- Motivation

- High Power Terahertz and mm-Wave Oscillators

- Wide Band Voltage Controlled Oscillators

- THz Varactor-less Scalable Standing Wave Radiator Array

- Terahertz Frequency Synthesizer

- High Gain mm-Wave and Terahertz Amplifier Design

65 **Challenges**

- CMOS VCOs (voltage controlled oscillator) normally use varactors
- CMOS Tunable sources above 150 GHz have < 4% of tuning range
- Two issues with varactors above 100 GHz :
  1. Poor quality factor (~2 @ 300GHz)
     ➢ Lossy resonator results in low power
  2. Large device parasitics in the resonator
     ➢ Limited frequency tuning range
- The locking range of the dividers is a strong function of injected power which is low at high frequencies.

$$Q = \frac{1}{\omega RC}$$

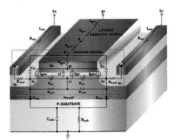

As mentioned low quality factor varactors are the major limitation behind wide band oscillators at mm wave and THz frequencies. In a synthesizer this also affects the locking range of the first divider because its input power is reduced due to the loss of the varactors.

66 **300 GHz Synthesizer**

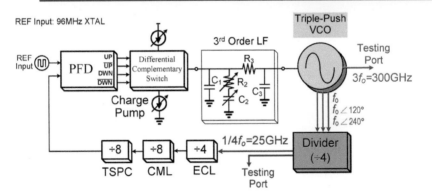

- The synthesizer employs a 3rd-order LF, a triple-push VCO, and a divide-by-1024
- The loop is locked to the VCO's fundamental at 100 GHz
- With three-phase injection, the divider achieves wider locking range

This is the block diagram of the 300 GHz synthesizer we implemented in a BICMOS process. The most challenging part of this system is the voltage controlled oscillator and the first divider. I will focus on these two blocks in this presentation.

Optimum Conditions   67

Assume output frequency of 300 GHz

	2-stage osc.	**3-stage osc.**	**4-stage osc.**
f_{osc}(GHz) / ϕ_{opt}	150 / 115°	100 / 130°	75 / 143°
Additional θ between each stage	65°	10°	53°
Number of transistors	2	3	4

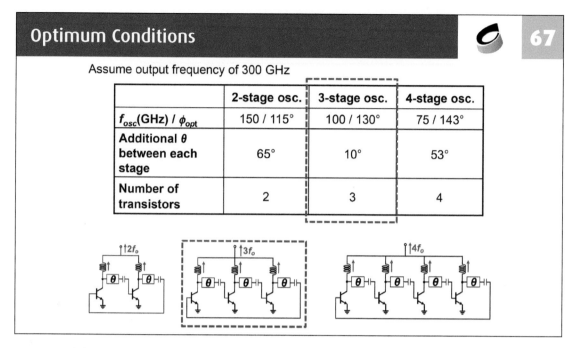

We used the same optimum conditions theory to choose the structure of the VCO. A 3 stage ring oscillator has the strongest oscillation and therefore the highest 3rd order harmonic as the output signal.

VCO with Colpitts-based Active Varactor   68

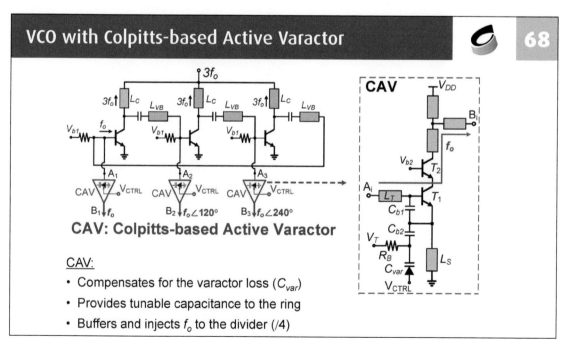

CAV: Colpitts-based Active Varactor

CAV:

- Compensates for the varactor loss (C_{var})
- Provides tunable capacitance to the ring
- Buffers and injects f_o to the divider (/4)

To avoid the loss of the varactors in the VCO we introduce a new block called Colpitts-based Active Varactor (CAV). The input impedance of this block provides a variable capacitor which is basically loss less. Moreover CAV behaves as the buffer that transfer the VCO signal to the first divider. This way the capacitive loading of the buffer on the VCO is eliminated.

69 ⊙ Divide by 4 Loop with 3-Phase Injection

- The free-running frequency of the divider loop is close to $1/4f_o$
- $3/4f_o$ generated from the amplifier A is mixed with the three-phase input signal
- The loop is locked to $1/4f_o$, i.e., frequency difference component

This is the block diagram of the divide by 4 frequency divider.

70 ⊙ Divide by 4 Loop with 3-Phase Injection

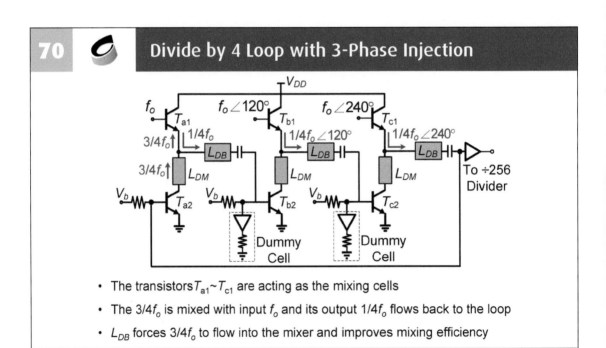

- The transistors $T_{a1} \sim T_{c1}$ are acting as the mixing cells
- The $3/4f_o$ is mixed with input f_o and its output $1/4f_o$ flows back to the loop
- L_{DB} forces $3/4f_o$ to flow into the mixer and improves mixing efficiency

The detailed circuit diagram of the frequency divider is shown here. The fundamental frequency of oscillation here is 25 GHz and the injected signal is 100 GHz. The 3 phase injection significantly increases the locking range of this divider.

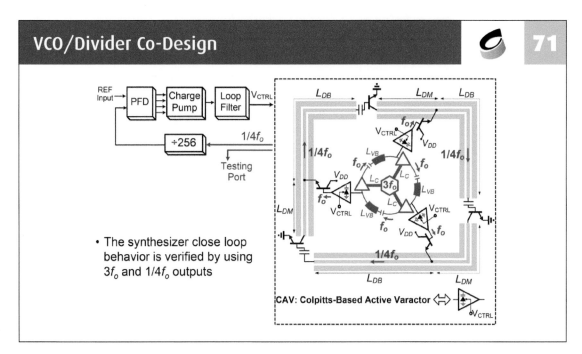

Because both VCO and divider have 3 stage structures, combining both layouts would be less complicated. The fig shows the shape of the actual layout on chip. The inner loop is the VCO which injects the signals to the outer loop which is the divider.

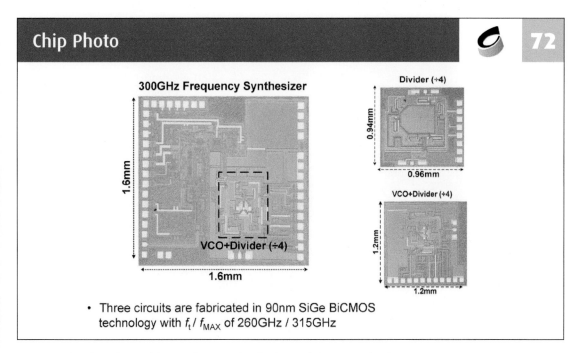

Here is the chip photo of the synthesizer, the divider, and the VCO + divider.

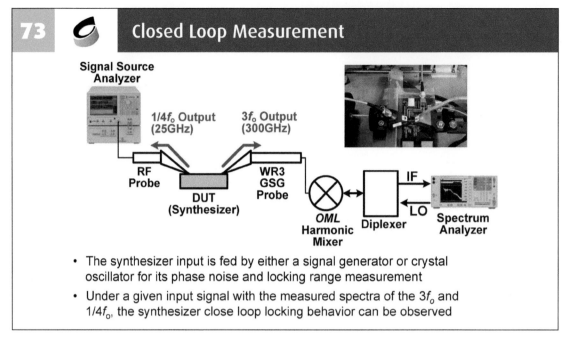

73   Closed Loop Measurement

- The synthesizer input is fed by either a signal generator or crystal oscillator for its phase noise and locking range measurement
- Under a given input signal with the measured spectra of the $3f_o$ and $1/4f_o$, the synthesizer close loop locking behavior can be observed

Harmonic mixer and power meters are used to detect the signal and characterize the synthesizer. The reference of the synthesizer is from both crystal and signal generator.

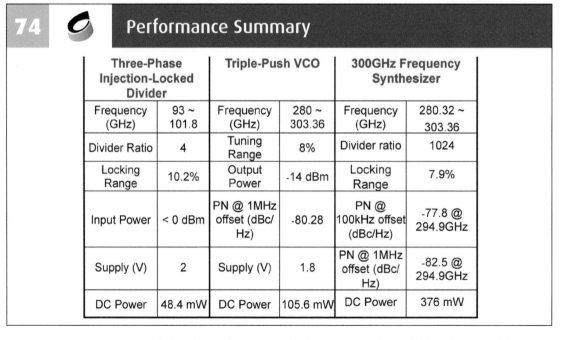

74   Performance Summary

Three-Phase Injection-Locked Divider		Triple-Push VCO		300GHz Frequency Synthesizer	
Frequency (GHz)	93 ~ 101.8	Frequency (GHz)	280 ~ 303.36	Frequency (GHz)	280.32 ~ 303.36
Divider Ratio	4	Tuning Range	8%	Divider ratio	1024
Locking Range	10.2%	Output Power	-14 dBm	Locking Range	7.9%
Input Power	< 0 dBm	PN @ 1MHz offset (dBc/Hz)	-80.28	PN @ 100kHz offset (dBc/Hz)	-77.8 @ 294.9GHz
Supply (V)	2	Supply (V)	1.8	PN @ 1MHz offset (dBc/Hz)	-82.5 @ 294.9GHz
DC Power	48.4 mW	DC Power	105.6 mW	DC Power	376 mW

Here is the performance table for the synthesizer and both VCO and 3-phase divider. The VCO achieves 8% tuning range which is almost equal to the locking range of the PLL.

Comparison Table 75

	This work	MTT-S 2011 [3]	JSSC 2011 [2]
Frequency (GHz)	280.32 ~ 303.36 (3rd)	300.76 ~ 301.12 (fund.)	160 ~ 169 (2nd)
Divider ratio	1024	10	128
Locking Range	7.9%	0.12%	5.5%
PN @ 100kHz/ 1MHz offset (dBc/ Hz)	-77.8 / -82.5 @ 294.9 GHz	-78 / -85 @ 300.96 GHz	-75 / -78 @ 163 GHz
DC Power (P_D)	376 mW	301.6 mW	1250 mW
FOM_T^* @ 100kHz/ 1MHz offset	-179.4 / -163.9 dBc/Hz	-144.4 / -131.36 dBc/Hz	-163.1 / -146.1 dBc/Hz
Technology (f_{max})	90nm SiGe BiCMOS (315 GHz)	InP HBT (800 GHz)	130nm SiGe BiCMOS (280 GHz)

$$* FOM_T = PN - 20\log\left(\frac{f_o}{\Delta f} \cdot \frac{Locking\ Range}{10}\right) + 10\log\left(\frac{P_D}{1mW}\right)$$

This work is the highest frequency synthesizer with the highest locking range in silicon and compound semiconductors.

Outline 76

- Motivation

- High Power Terahertz and mm-Wave Oscillators

- Wide Band Voltage Controlled Oscillators

- THz Varactor-less Scalable Standing Wave Radiator Array

- Terahertz Frequency Synthesizer

- High Gain mm-Wave and Terahertz Amplifier Design

77 **CMOS mm-Wave & THz Amplifiers**

- Signal amplification is challenging in CMOS:

  - CMOS scaling is reaching its limit.

  - Operation frequency of these systems is close to the maximum oscillation frequency (f_{max}) of the transistors.

  - Maximum available gain (G_{ma}) of the transistors drops below useful level for most applications.

  - PAE drops as the gain drops at high frequencies.

  > **We need to boost G_{ma} to its maximum possible value.**

S ignal amplification is challenging at mm wave and THz frequencies in CMOS. Because of digital/Rf integration CMOS is attractive for these systems. However CMOS scaling is reaching its limit and the operation frequency of these systems is close to the maximum oscillation frequency of the CMOS transistors. This results in Maximum available gain (Gma) of the transistors to significantly drop and become useless for most applications. PAE also drops as the gain drops at this region. The solution is to boost the Gma to its maximum theoretical value.

78 **Maximum Available Gain (G_{ma})**

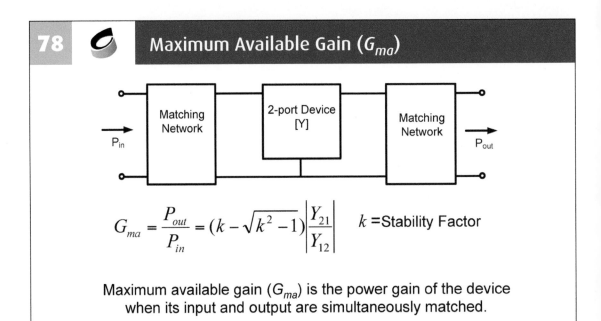

$$G_{ma} = \frac{P_{out}}{P_{in}} = (k - \sqrt{k^2 - 1})\left|\frac{Y_{21}}{Y_{12}}\right| \qquad k = \text{Stability Factor}$$

Maximum available gain (G_{ma}) is the power gain of the device when its input and output are simultaneously matched.

B ased on the definition maximum available gain (Gma) is the power gain of a 2-port device when the input and output are simultaneously matched. As you can see this gain is a function of the 2-port device only.

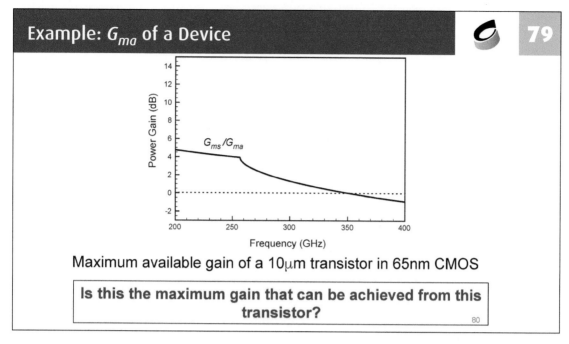

Example: G_{ma} of a Device

Maximum available gain of a 10μm transistor in 65nm CMOS

Is this the maximum gain that can be achieved from this transistor?

Just to give you sense of the numbers, in a 65 nm CMOS process the maximum available gain is around 2 dB at 270 GHz. The question is that is this the maximum gain that can be achieved from this specific transistor?

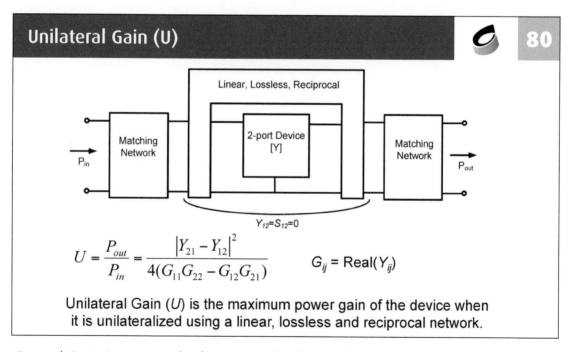

Unilateral Gain (U)

$Y_{12}=S_{12}=0$

$$U = \frac{P_{out}}{P_{in}} = \frac{|Y_{21} - Y_{12}|^2}{4(G_{11}G_{22} - G_{12}G_{21})} \qquad G_{ij} = \text{Real}(Y_{ij})$$

Unilateral Gain (U) is the maximum power gain of the device when it is unilateralized using a linear, lossless and reciprocal network.

One solution to increase Gma has been to neutralize the gate-drain capacitance of the transistor, make it unilateral and boost the available gain. The maximum gain that can be achieved by unilateralization is called U. the way to reach this gain is to embed the device in a linear lossless reciprocal network to create a zero reverse gain and simultaneously match the input and output of the resulting network. As I will mention later U is only a function of the 2-port device we started with. Is this the max gain that a device can achieve?

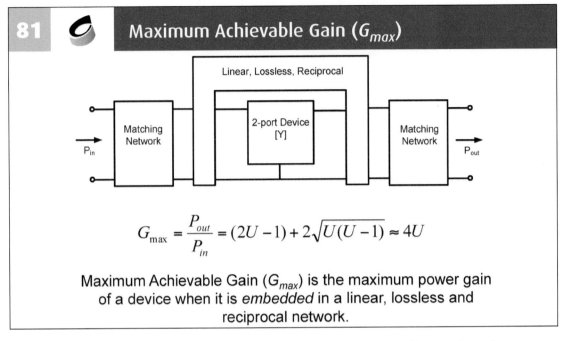

The answer is no. There is another gain definition which I call maximum achievable gain. This is the maximum power gain that can be achieved from a 2-port device when it is embedded in a linear, lossless and reciprocal network. This gain is a function of U and hence is only a function of the 2-port device.

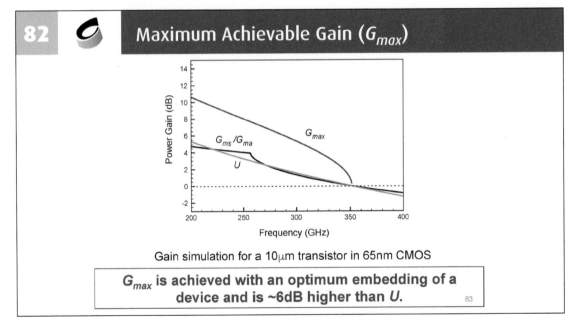

This is the plot of the unilateral gain and Gmax for the same device in 65 nm CMOS. It is clear that U does not improve the gain compared to the maximum available gain of the device at frequencies close to the fmax. This is because the gain is mainly limited by the loss of the transistor rather than the power leak from output to input or vice versa. Gmax is achieved by the optimum embedding and is 6 dB higher than unilateral gain, U. at 270 GHz Gmax is around 7 dB compared to U and Gma which is 2 dB at the same frequency. Just to emphasize, this is the maximum gain that can be achieved while the amplifier is unconditionally stable.

Power Flow of a Device

 83

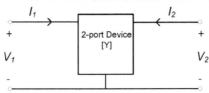

$$P = V_1^* I_1 + V_2^* I_2 \qquad\qquad A = \left|\frac{V_2}{V_1}\right| \ \& \ \varphi = \angle(\frac{V_2}{V_1})$$

$$P_R = P_{out} - P_{in}$$

$$= |V_1|^2 \{-(G_{11} + A^2 G_{22}) - A|Y_{12} + Y_{21}^*|\cos(\angle(Y_{12} + Y_{21}^*) + \varphi)\}$$

$$\boxed{\frac{P_{out}}{P_{in}} = 1 + \frac{|V_1|^2}{P_{in}}\{-(G_{11} + A^2 G_{22}) - A|Y_{12} + Y_{21}^*|\cos(\angle(Y_{12} + Y_{21}^*) + \varphi)\}}$$

I n order to find the optimum conditions to reach Gmax we again look into the fundamental power flow of the transistor. As shown above we can find the power gain of the transistor as a function of the voltage swing at the input, Y parameters of the device, and A and phi.

Optimum Conditions

 84

Assuming most of the input power flows from source to port 1 of the device and most of the output power flows to the load:

$$Gain \cong 1 + \frac{2}{G_{11}}\{-(G_{11} + A^2 G_{22}) - A|Y_{12} + Y_{21}^*|\cos(\angle(Y_{12} + Y_{21}^*) + \varphi)\}$$

$$\boxed{A_{opt} = \frac{|Y_{12} + Y_{21}^*|}{2G_{22}} \qquad \varphi_{opt} = (2k+1)\pi - \angle(Y_{12} + Y_{21}^*)}$$

Gain is maximized by providing optimum voltage gain and phase conditions for the device.

B y making an assumption we can find the power gain of the device as a function of Y parameters, A and phi only. With this we can find the optimum conditions that results in the maximum power gain. This power gain would be G_{max}.

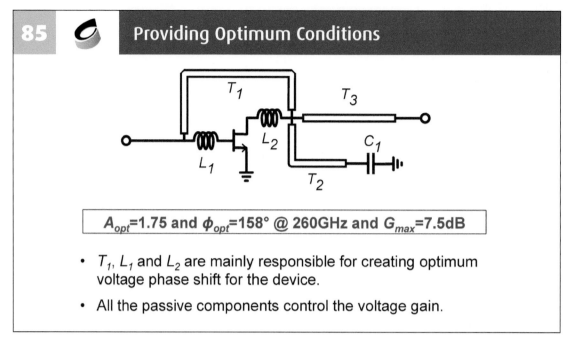

85 — Providing Optimum Conditions

A_{opt}=1.75 and ϕ_{opt}=158° @ 260GHz and G_{max}=7.5dB

- T_1, L_1 and L_2 are mainly responsible for creating optimum voltage phase shift for the device.

- All the passive components control the voltage gain.

The simulated optimum conditions are found and the passive components around the device is designed to reach these numbers.

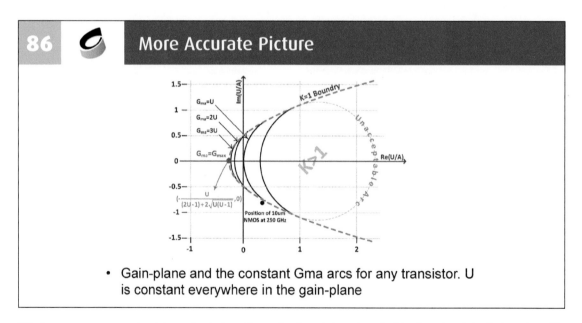

86 — More Accurate Picture

- Gain-plane and the constant Gma arcs for any transistor. U is constant everywhere in the gain-plane

To have a more accurate picture of gain boosting gain plane is introduced. This plane is originally introduced in [1]. Any device or network can basically be presented as a point at a specific frequency on this plane. X and Y axis are Re(U/A) and Imag(U/A) respectively. A is defined as the max stable gain S21/S12. Any Gma value has an arc in this plane and Gmax is represented by a dot as shown here. A 10 um NMOS transistor at 260 GHz is also shown in this plane. The objective of the designer is to move from the coordinate of the devise to the coordinate of Gmax or any desired gain.

[1] A. Singhakowinta and A. Boothroyd, "Gain Capability of Two-port Amplifiers," International Journal of Electronics, vol. 21, no. 6, 1966, pp. 549–560.

Parallel Embedding   87

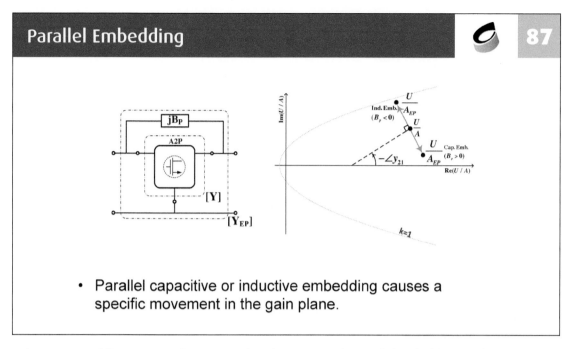

- Parallel capacitive or inductive embedding causes a specific movement in the gain plane.

There are two different ways of moving in this plane. One is the parallel embedding which results in the movement shown in the fig. capacitive and inductive embeddings have opposite directions.

Shunt Embedding   88

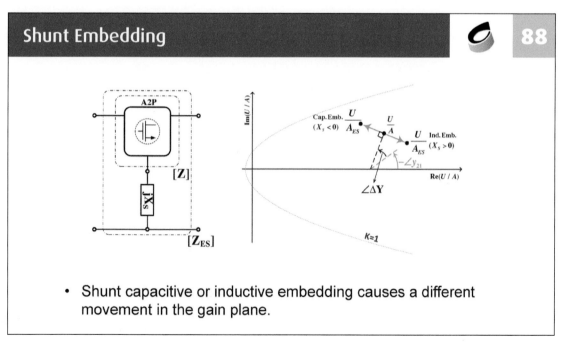

- Shunt capacitive or inductive embedding causes a different movement in the gain plane.

Shunt embedding has a different movement with a different angle in this plane.

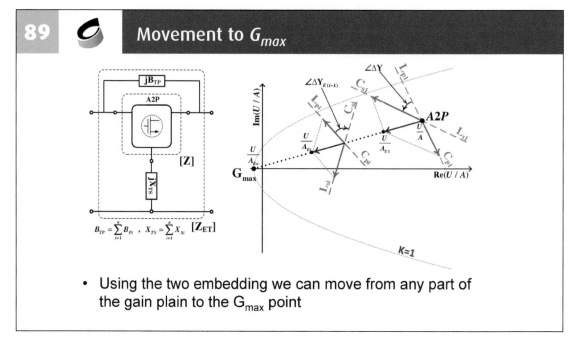

- Using the two embedding we can move from any part of the gain plain to the G_{max} point

Summation of both of these embedding results in a movement that can be engineered to have any direction and distant. For example we can move from any device in the plane to G_{max} and achieve the highest power gain possible.

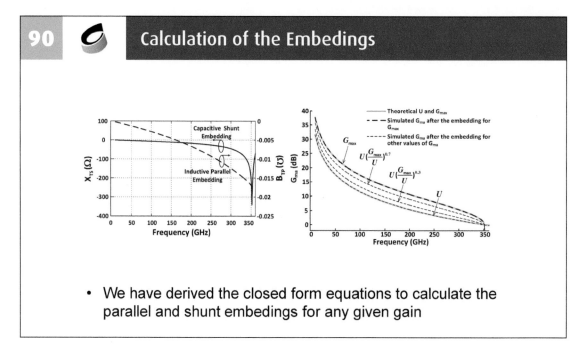

- We have derived the closed form equations to calculate the parallel and shunt embedings for any given gain

We have calculated the exact component values as a function of the Y parameters of the device ad the desired power gain. If the device is given we can readily find the optimum embedding to achieve the highest power gain.

Circuit Simulation and Verification   91

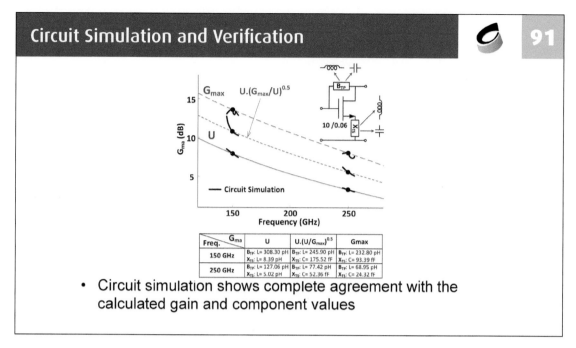

- Circuit simulation shows complete agreement with the calculated gain and component values

Circuit simulations verifies the equation and derivations for this optimum embedding.

Pre Embedding   92

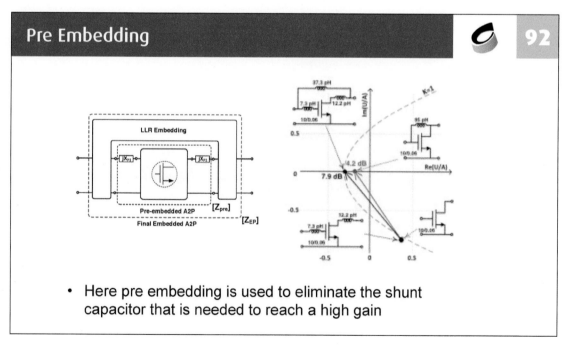

- Here pre embedding is used to eliminate the shunt capacitor that is needed to reach a high gain

Pre embedding can also be used to simplify the structure and reduce the loss associated with the embedding.

93 Implemented Amplifier

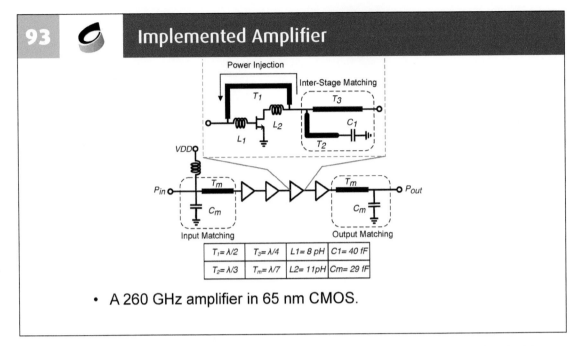

$T_1= \lambda/2$	$T_3= \lambda/4$	$L1= 8 \text{ pH}$	$C1= 40 \text{ fF}$
$T_2= \lambda/3$	$T_m= \lambda/7$	$L2= 11 \text{pH}$	$Cm= 29 \text{ fF}$

- A 260 GHz amplifier in 65 nm CMOS.

The circuit schematic of the implemented amplifier is shown here. Pre-embedding and parallel embedding are used to boost the gain at 260 GHz in a 65 nm CMOS process.

94 Chip Photo

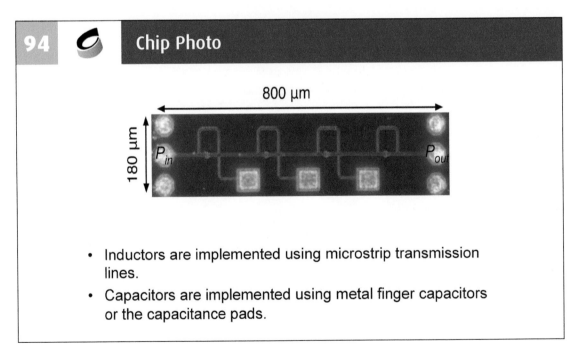

- Inductors are implemented using microstrip transmission lines.
- Capacitors are implemented using metal finger capacitors or the capacitance pads.

Here is the chip photo of the amplifier.

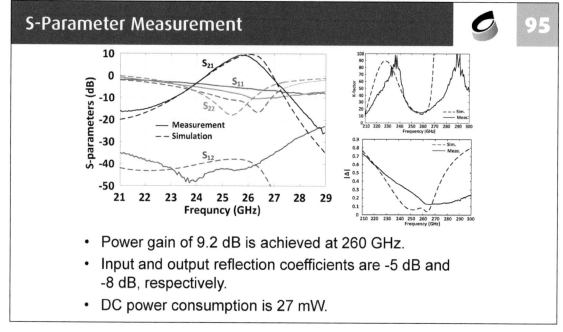

S-Parameter Measurement

- Power gain of 9.2 dB is achieved at 260 GHz.
- Input and output reflection coefficients are -5 dB and -8 dB, respectively.
- DC power consumption is 27 mW.

M easurement shows 9.2 dB gain at 260 GHz while the amplifier is unconditionally stable.

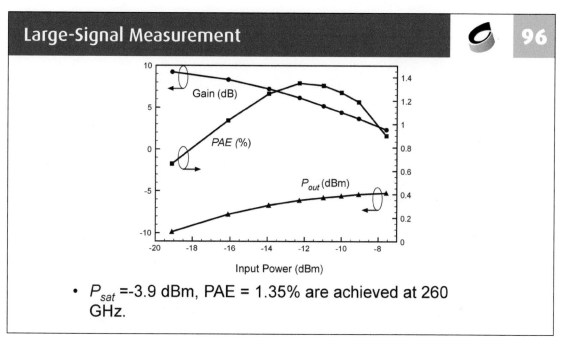

Large-Signal Measurement

- P_{sat} =-3.9 dBm, PAE = 1.35% are achieved at 260 GHz.

S aturation power of -3.9 dBm is achieved at 260 GHz.

97

Comparison

Ref.	Tech.	f_{max} (GHz)	Freq. (GHz)	Gain (dB)	3-dB BW (GHz)	VDD (V)	P_{1-dB} (dBm)	P_{sat} (dBm)	Peak PAE (%)	P_{DC} (mW)	Topology	Area (mm^2)
[28]	0.13 µm SiGe HBT	435	220	16	28	3.6	N/A	N/A	N/A	144	3 Diff. Cascode Stages	0.45
[29]	250 nm InP DHBT	550	324	4.8	3 (2-dB)	1.4	N/A	1.1	0.6	16.8	1 CB stage	0.12
[30]	50 nm InP HEMT	1200	340	15	25^1	2.2	N/A	10	N/A	294.8	4 CE stages	0.49
[31]	40 nm CMOS	275	213.5	10.5	13	0.8	-7.2	-3.2	0.75	42.3	9 CS stages	0.013
[32]	50 nm InP HEMT	N/A	220	10.4	30	2.2	18	N/A	3.7	2165	4 CS stages 8 power comb.	0.96
[33]	250 nm InP HEMT	700	220	8.9	36	2	N/A	18.1	N/A	1018	1 CE/ 1 CB 8 power comb.	0.99
[34]	32 nm SOI CMOS	320	210	15	14	1	2.7	4.6	6	40	3 Neutralized diff. stages	0.06
This work	65 nm CMOS	352 GHz	257	9.2	12.2	0.8	-8	-5.5	1.3	13.7	4 CS stages	0.14
This work	65 nm CMOS	352 GHz	257	8.5	12.2	1	-8	-3.9	0.8	27.6	4 CS stages	0.14

[1] Estimation from the reported S-parameters.

- The highest frequency silicon-based amplifier

Here is the comparison table with the state of the art. This work is the highest frequency amplifier in silicon. The P_{sat} of the amplifier is higher than any silicon-based amplifier beyond 220GHz.

98

Conclusion

- To implement terahertz systems, we need to find and reach the fundamental frequency/power limits of various blocks.
- We showed a few key blocks of the system on standard CMOS process:

  - Oscillator and amplification close to f_{max}.
  - High power generation much above f_{max}.
  - Wide tuning range signal source and synthesizer.

- The proposed circuits can be implemented in *any* process such as InP or GaAs for even higher frequency/power.

In order to push the frequency in conventional silicon process we need to understand the limits and find innovative methods to reach them. We have shown several building blocks at mm wave and THz frequencies that does the same and achieve record breaking performances.

CHAPTER 5

Equalization and A/D Conversion for High-Speed Links

Boris Murmann

Stanford University, USA

Ryan Boesch

Raytheon Space and Airborne Systems, USA

Kevin Zheng

Stanford University, USA

As modern electrical and optical communication systems transition toward advanced modulation schemes, there exists a pressing need for power efficient A/D converters operating at tens of gigasamples per second. Within this context, this chapter will cover circuit- and architecture-level design techniques for the front-end circuitry of ADC-based links. The first part will focus time-interleaved data conversion using SAR and flash topologies. The second part will look into the opportunity of relaxing the ADC requirements using analog pre-equalization.

Wireline Trend

Consistent upward trend in data rates for all standards due to increasing computing and communication requirements

25-28GT/s already deployed; next-gen reaches 56-64GT/s/Lane

>100GT/s/Lane seems inevitable even for electrical interconnects

› Transferring a 1GB size movie file at this speed takes less than 1sec!

› Keeping up with the traditional bit error rate (BER) of 10^{-12} is difficult
  → Forward error correction (FEC) is now built into most standards

Need more levels since raw BW is not keeping up: PAM2 → PAM4 → ...

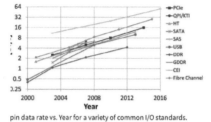

pin data rate vs. Year for a variety of common I/O standards.

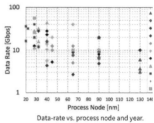

Data-rate vs. process node and year.

In the era of connectivity, wireline I/O has been a key technology underpinning the aggressive performance scaling of computer and communication systems. All standards, ranging from electrical to optical, long haul to short reach interconnects, have increased their aggregate bandwidth requirement at a rate of about 2.5x every 2 years. This trend drives the usage of multi-level modulation schemes, such as PAM4. As a result, ADC-DSP based links are gaining more attention and are now heavily investigated. ADC links also take advantage of the intrinsic bandwidth and area improvements from process technology scaling.

Link Architectures

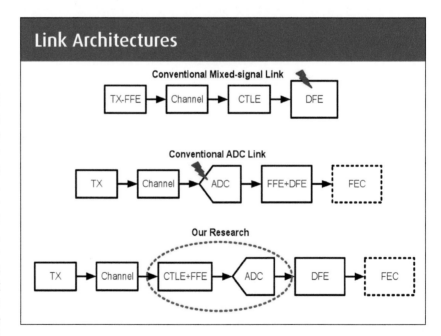

It is becoming very difficult for conventional mixed signal links (top drawing) to meet the bandwidth and performance requirements of next-gen systems. Transmitter pre-emphasis is limited by a peak power constraint. A continuous time linear equalizer (CTLE) is intrinsically blind about cursors and operates in frequency domain. The decision mixed-signal feedback equalizer (DFE) sees additional timing and scalability challenges. On the other hand, as equalization is moved into the digital domain for ADC-based links, the ADC needs to be very fast, have reasonably high resolution and yet be very power efficient. We have studied a new architecture that leverages an analog equalizer in front of the ADC to relax this challenge.

Outline

Specifying the receiver front-end
› Resolution requirements
› Impact of nonidealities
› Benefits of adding an analog FFE

Rx FFE design
› Prior art
› Inverter-based design
› Experimental results

Rx ADC design
› Performance trends
› Time interleaving
› Flash-based design
› SAR-based design

Our discussion has three main parts: (1) Specifying the receiver front-end, (2) a receive-FFE design example, and (3) a summary on the state-of-the art in ADC implementation for high-speed links.

Lost Opportunities...

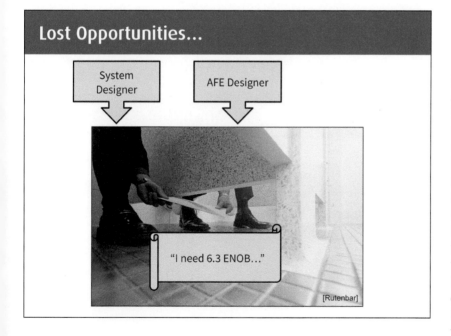

Starting with specifications for ADC-based links, a prevalent problem in the design community is that the interface between the system designer and circuit engineer relies on over-simplified metrics (such as ENOB – effective number of bits) that can lead to significant overdesign. Since overdesign translates into a loss in power efficiency, this is clearly an important issue. We therefore dedicate the first part of our discussion entirely to the derivation of proper circuit specifications.

5

In order to understand how ADCs affect the overall link performance and how they should be specified, we have to break down the individual nonidealities and consider each one separately. In this framework, the notion of ENOB (effective number of bits) is invalid, since it combines all noise sources by power addition without acknowledging their different natures. In subsequent parts, we will investigate the different impairments in an ADC link as illustrated in the slide.

Specifying the Receiver Front-End

TX → Channel → ADC → FFE+DFE → FEC

TX → Channel → CTLE+FFE → ADC → DFE → FEC

1. Input signal statistics?
2. Nonlinearity?
3. How do these blocks help?
4. What are the tradeoffs?

1. How many bits?
2. INL/DNL?
3. Full scale clipping?
4. Timing jitter/skew?

1. Impact of DSP equalizers on ADC noise?

6

The bit error rate follows from integrating the "tails" of the noise probability density functions that overlap beyond the decision boundary. The shown example is for PAM2, but the concept extends naturally to other modulation schemes.

Bit Error Rate (BER) – Main Idea

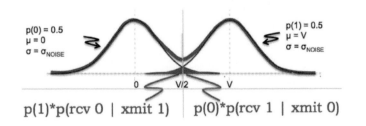

$p(0) = 0.5$
$\mu = 0$
$\sigma = \sigma_{NOISE}$

$p(1) = 0.5$
$\mu = V$
$\sigma = \sigma_{NOISE}$

0   V/2   V

$p(1)*p(\text{rcv } 0 \mid \text{xmit } 1)$ $p(0)*p(\text{rcv } 1 \mid \text{xmit } 0)$

- Shown for PAM2, but same idea applies to higher order modulations

BER Calculation with Quantization Noise

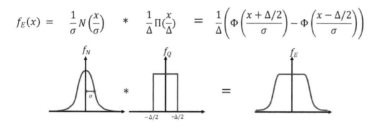

$$f_E(x) = \frac{1}{\sigma}N\left(\frac{x}{\sigma}\right) \; * \; \frac{1}{\Delta}\Pi\left(\frac{x}{\Delta}\right) \; = \; \frac{1}{\Delta}\left(\Phi\left(\frac{x+\Delta/2}{\sigma}\right) - \Phi\left(\frac{x-\Delta/2}{\sigma}\right)\right)$$

Definitions:

$$N(x) = \frac{1}{\sqrt{2\pi}}e^{-x^2}$$

$$\Pi(x) = \begin{cases} 1, |x| < 1/2 \\ 0, |x| \geq 1/2 \end{cases}$$

$$\Phi(x) = \int_{-\infty}^{x} N(u)\,du$$

- The BER should be calculated by integrating the tail of $f_E(x)$ up until the eye opening height $(-\mu)$
- Cannot simply add ADC quantization noise power and use the Φ function approach!

✔️ $BER = \int_{-\infty}^{-\mu} f_E(x)\,dx$ ❌ $BER = \Phi\left(\frac{\mu}{\sqrt{\Delta^2/12 + \sigma^2}}\right)$

and this yields a difference between integrals of the Gaussian function, since a uniform distribution is a difference of step functions. The correct BER estimation is done by integrating the tail of the new PDF, instead of simply using the Gaussian cumulative distribution function with the added noise power from quantization. It is important to note here that since a uniform PDF is a bounded noise, its effect on BER degradation intuitively should be less significant than that of a noise source that has a long tail.

We assume that the system's thermal noise and quantization noise are independent (a good assumption for the system in question). The overall PDF follows by convoluting the two individual PDFs

Wrong BER Calculation Leads to Overdesign

- In the example below 3-bit quantization within the eye opening provides constitutes a feasible design point
- 2-bit quantization may still be feasible for systems with FEC

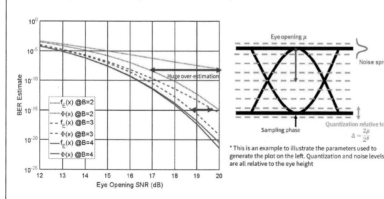

* This is an example to illustrate the parameters used to generate the plot on the left. Quantization and noise levels are all relative to the eye height

nal eye opening (which is proportional to the channel main cursor value). The illustration on the right shows the parameters used for the PDFs. With 4-bit quantization within the eye, the quantization noise is essentially negligible, since the thermal noise dominates. In the extreme case of 2-bit quantization, the wrong method gives a dramatic BER overestimation. A

The plot on the left gives a feel for how a wrong BER estimation can lead to overdesign. We set the quantizer resolution level with respect to the fi-

3-bit quantizer for the eye opening looks like a good choice in this example.

First-Order ADC Resolution Requirement

A first-order resolution requirement can now be formulated. The most important three parameters are the base eye opening, the quantizer resolution, the number of PAM modulation levels, and the channel impairment. The channel impairment comes from the fact that not the whole ADC full scale range is used to quantize the eye opening, but ISI is also taking up some signal swing. We will discuss the channel degradation in more detail later.

$$B = B_{Eye} + \log_2(\text{PAM} - 1) + B_{Channel}$$

Pick base eye opening resolution B_{Eye}
> In this example, use 3 bits with margin in mind

Add resolution due to PAM modulation since there will be more eyes to be quantized
> For PAM4, add $\log_2(4 - 1) \approx 1.6$ bits

Add extra range due to channel
> For typical channels with bad ISI, it can lead to >1.5 bits degradation

$$B = 3 + 1.6 + 1.5 \approx 6$$

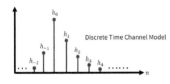

Discrete Time Channel Model

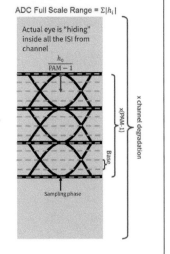

ADC Full Scale Range = $\Sigma|h_i|$

Actual eye is "hiding" inside all the ISI from channel

$\dfrac{h_0}{\text{PAM} - 1}$

Sampling phase

For this particular example, 6 bits would be a good design starting point for PAM4 modulation.

Analog Front End (AFE) Nonlinearity

We developed a framework to think about quantization noise in the PDF domain, and now we will extend it to AFE nonlinearity as well. We will start with static nonlinearity, specifically 3rd order compression. Assuming that the equalizer can take care of the linear part of the signal chain completely, the 3rd order term now simply becomes another "noise" source, independent from the other noises. It is important to note that nonlinearities in links have to be treated differently from the usual sinusoidal analysis, since the input signal is wideband and it also has a different amplitude distribution.

- The AFE's nonlinearity affects any link in general (not specific to ADC links)
- Higher PAM modulation scheme is more sensitive to nonlinearity
- Conventional nonlinearity specs for sinusoidal signals won't work

$$G_{NL}(x) = x - cx^3$$

$$\text{error} = q + \sigma - cx^3$$

Can be treated as an independent error source

AFE Static Nonlinearity Analysis

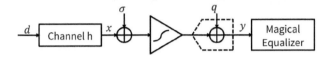

- Given input PDF $f_X(x)$, what is f_Y when $y = cx^3$?

$$F_X(x) = F_X\left(\sqrt[3]{\left(\frac{y}{c}\right)}\right) \Rightarrow f_Y(y) = \frac{d}{dy} F_X\left(\sqrt[3]{\left(\frac{y}{c}\right)}\right) = \frac{1}{3c}\left(\frac{y}{c}\right)^{-\frac{2}{3}} f_X\left(\sqrt[3]{\left(\frac{y}{c}\right)}\right)$$

- We need to separate out the nonlinearity error PDFs for different transmit data. Therefore, conditional PDFs must be used:

$$f_{NL|d=D}(y) = \frac{1}{3c}\left(\frac{y}{c}\right)^{-\frac{2}{3}} f_{X|d=D}\left(\sqrt[3]{\left(\frac{y}{c}\right)}\right)$$

Knowing the input signal PDF, the exact PDF for the nonlinear term can be found. However, we need to separate out the conditional PDFs for different transmit data. The input PDF of data = 1 and that of data = 1/3 will yield different distorted PDFs. This is due to the fact that each conditional input PDF operates on different parts of the nonlinear DC transfer curves.

AFE Static Nonlinearity Analysis, continued

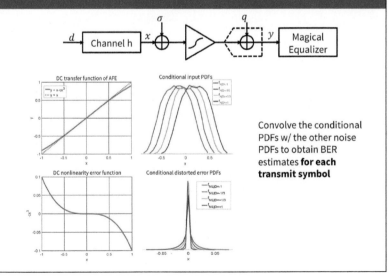

Convolve the conditional PDFs w/ the other noise PDFs to obtain BER estimates **for each transmit symbol**

Due to heavy ISI and randomness of data, the conditional input PDFs (upper right) look very similar, but are centered at different input values. Therefore, when each PDF projects onto the corresponding portions of the nonlinear error function (lower left), the resulting conditional nonlinearity error PDFs are very different (lower right). For example, the PDF of d=1 gives a much larger left tail than d=1/3. This will directly result in a larger BER when the transmit data is 1.

13

In a real circuit, there can be multiple sources of nonlinearity and we have already established the fact that a simple sine wave test can't tell the whole story. Shown on this slide is a systematic way to directly simulate AFE nonlinearity and correlate it to BER. We simulate the AFE with a Pseudo-Random Bit Sequence (PRBS), including the channel. Then, in a system modeling tool like MATLAB, we can extract the linear component of the signal path using LMS system identification (SID). The residual errors will be from nonlinearity.

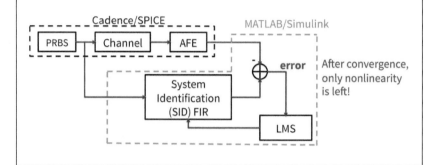

Nonlinearity Simulation for Links

- Real circuits also have dynamic nonlinearities in addition to static terms
  › The dynamic potion is harder to model
- The whole signal chain can include CTLE, VGA, and ADC S/H

14

In this example, a MATLAB-generated channel output is sent through a SPICE circuit model to obtain the output transient signal. After identifying the system using the LMS algorithm, the residual error captures the nonlinear error. The accuracy of nonlinear error PDF depends on the length of PRBS signal used. Normally, PRBS15 can be used for a reasonable tradeoff between simulation time and sufficient signal statistics. With this obtained information, we can compute directly how this nonlinearity affects the BER performance.

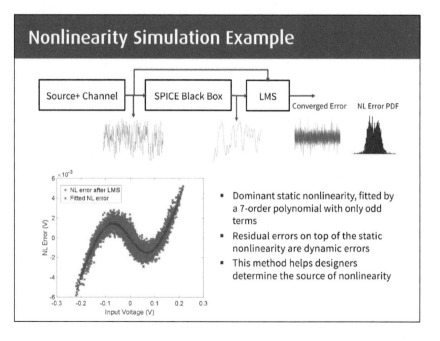

Nonlinearity Simulation Example

- Dominant static nonlinearity, fitted by a 7-order polynomial with only odd terms
- Residual errors on top of the static nonlinearity are dynamic errors
- This method helps designers determine the source of nonlinearity

Nonlinearity Simulation Example, Continued

 15

- By convolving the obtained nonlinearity PDFs with other BER estimation PDFs, we can obtain the following curves
- In this example, peak NL error is ~1/3 of the eye opening
- With the DC nonlinearity error alone, a sine wave test gives a 3rd order harmonic of -36.5dBc @full scale of the input signal, but this doesn't say anything about the final BER performance

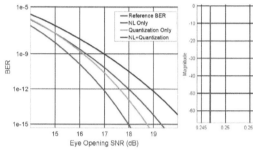 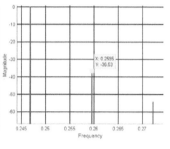

With the extracted NL PDF, we can visually see how it changes the BER curve under different

thermal noise environment. When only thermal noise is present, we obtain the reference BER curve where it intersects 1e-12 with 17dB SNR. We have seen the degradation due to 3-bit eye opening quantization as shown by the yellow curve. The example NL errors give similar degradation in this case. The combined noise reduces noise margin by ~2dB @BER=1e-12. This degradation is even lower if the raw BER of interest is larger.

Impact of DNL

 16

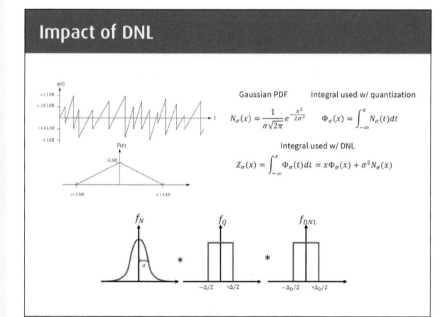

When INL has polynomial shape, it can be treated as part of static nonlinearity. The analysis for AFE nonlinearity applies. On the other hand,

DNL directly affects the shape of quantization error function, thus the quantization error PDF will be different. If we assume approximately another uniform distribution for DNL PDF, the error PDF will be convolving another uniform distribution on top of the original with quantization alone (drawing by Ion Opris). We will call the integral of Φ the Z function. Not surprisingly, there is also a closed form for the Z function. We will analyze the effect of DNL on BER using this method.

17

We already know that convolving with $\pm\Delta/2$ uniform PDF for quantization will give a difference of Φ functions. By convolving again with $\pm\Delta_d/2$ uniform PDF for DNL, each Φ function will become a difference of Z functions. However, this closed form is already starting to grow out of hand. We investigate how some simple approximations can pro-

Impact of DNL, Continued

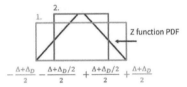

$$f_E = \frac{1}{\Delta}\left(\Phi_\sigma\left(x+\frac{\Delta}{2}\right) - \Phi_\sigma\left(x-\frac{\Delta}{2}\right)\right)$$

$$f_{E+DNL} = \frac{1}{\Delta\cdot\Delta_D}\left\{\left(Z_\sigma\left(x+\frac{\Delta+\Delta_D}{2}\right) - Z_\sigma\left(x+\frac{\Delta-\Delta_D}{2}\right)\right) - \left(Z_\sigma\left(x-\frac{\Delta-\Delta_D}{2}\right) - Z_\sigma\left(x-\frac{\Delta+\Delta_D}{2}\right)\right)\right\}$$

- Approximations:
  - › 1. Add Δ_D to the nominal Δ as the new LSB in f_E to estimate.
  - › 2. Add $\Delta_D/2$ to the nominal Δ as the new LSB in f_E to estimate.

vide good enough estimations. We either approximate the trapezoid with a rectangle of the same base width or a rectangle with the same height, both with the same area since they are all PDFs. Then the original single uniform PDF estimation function can be used.

18

The actual Z function estimation is bounded by the two approximation methods. The equal height method provides a closer estimation while the equal base method gives slight over-estimation. When the eye opening quantization resolution is high, the effect of DNL is also negligible just like the quantization itself. We obtain a rule of thumb for DNL speci-

DNL Effects on BER

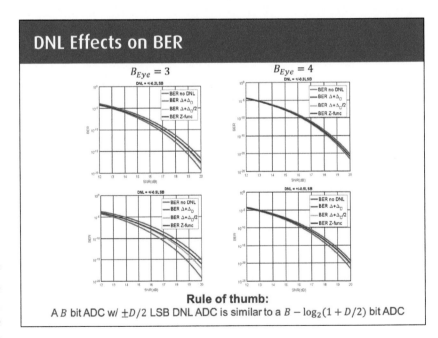

Rule of thumb:
A B bit ADC w/ $\pm D/2$ LSB DNL ADC is similar to a $B - \log_2(1 + D/2)$ bit ADC

fication as shown in the slide. For a typical $\pm0.5 LSB$ DNL, add ~50% more levels (applicable to Flash) if quantization noise dominates in system.

Full Scale Range Clipping

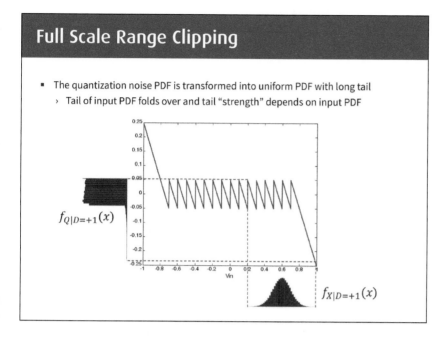

19

- The quantization noise PDF is transformed into uniform PDF with long tail
  - Tail of input PDF folds over and tail "strength" depends on input PDF

$f_{Q|D=+1}(x)$

$f_{X|D=+1}(x)$

Real ADCs all have finite full-scale ranges. When an ADC is clipped, conventional wisdom says something bad will happen. ADC FSR clipping can be translated as an extended linear function in the quantization error function. When an input PDF clips the ADC, a part of the PDF folds directly into the quantization error PDF. As a result, the quantization PDF is no longer bounded by half LSBs, but is a function of the input PDF. The subtle tradeoff here is that the main cursor strength becomes larger relative to the nominal LSB, therefore an optimal clipping ratio should exist.

Full Scale Range Clipping, continued

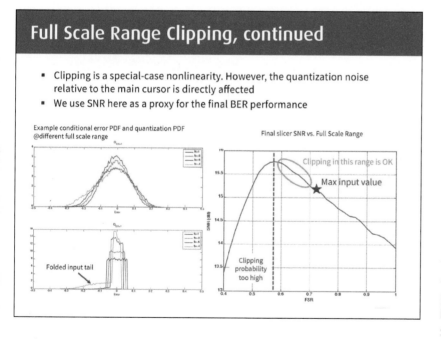

20

- Clipping is a special-case nonlinearity. However, the quantization noise relative to the main cursor is directly affected
- We use SNR here as a proxy for the final BER performance

Example conditional error PDF and quantization PDF @different full scale range

Folded input tail

Final slicer SNR vs. Full Scale Range

Clipping in this range is OK

Max input value

Clipping probability too high

In this example, the FSR of the ADC is tuned so that different clipping ratios are achieved. When the quantization error PDF (bottom left) is extracted, we see more tail folding from the input PDF for smaller FSR. Interestingly, there is an optimal FSR that gives maximum SNR. After that point, the SNR falls off the cliff since the equalizer adaptation fails. We conclude that there is a region of clipping ratio that can provide acceptable if not better performance than simply setting the FSR to be the maximum input value. In other words, the dynamic range of the ADC has to be fully utilized.

21

Another important nonideality is ADC sampling jitter. Available textbook analysis often only consider the impact of jitter for sinusoidal signals. As shown on this slide, assuming sinusoids at fs/2 leads to very low SNR numbers in presence of practical amounts of jitter.

Impact of Jitter

- Textbook jitter analysis typically assumes a sinusoidal input at $f_s/2$, which is far too pessimistic for a communication link

$$SNR_t = \frac{P_{sig}}{P_t} = \frac{\frac{1}{2}A^2}{\frac{1}{2}A^2 \omega_{sin}^2 \cdot \sigma_t^2} = \frac{1}{\omega_{sin}^2 \cdot \sigma_t^2}$$

$$\alpha_t = \omega_{sin}^2$$

Jitter conversion factor

Example: 10GHz sinusoid, 1ps$_{rms}$ jitter

$$\alpha_t \cdot \sigma_t^2 = (2\pi 10GHz)^2 1ps^2 = \frac{3.9 \cdot 10^{-3}}{ps^2} \cdot 1ps^2 = -24dB$$

(ouch!)

22

A better way to estimate the impact of jitter is to consider the signal's autocorrelation function. A fast moving signal has a narrow autocorrelation function, while a slowly moving signal shows a very gradual roll-off in its autocorrelation.

Improved Jitter Analysis

- Da Dalt (TCAS1, 9/2002) showed that α_t follows from the curvature of the autocorrelation function (for arbitrary signals with second order differentiable autocorrelation)

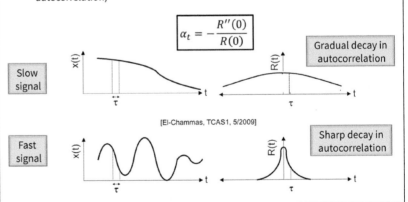

$$\alpha_t = -\frac{R''(0)}{R(0)}$$

[El-Chammas, TCAS1, 5/2009]

Example (1)

23

- Neglecting the frequency response of TX and RX for simplicity

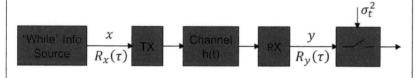

$$R_y(\tau) = R_x(\tau) * h(t) * h(-t)$$

$$R_y(\tau) = \delta(\tau)\sigma_x^2 * h(t) * h(-t) = \sigma_x^2 \cdot \underbrace{[h(t) * h(-t)]}_{w(\tau)}$$

$$\boxed{\alpha_t = -\frac{R_y''(0)}{R_y(0)} = -\frac{w''(0)}{w(0)}}$$

The autocorrelation at the sampler input can be found by convolving the input autocorrelation (usually a delta function) with the input response of the channel. See Da Dalt TCAS1, 9/2002, and El-Chammas, TCAS1, 5/2009 for further details. The conversion factor for jitter follows from the curvature of the autocorrelation, as expressed in the final result.

Example (2)

24

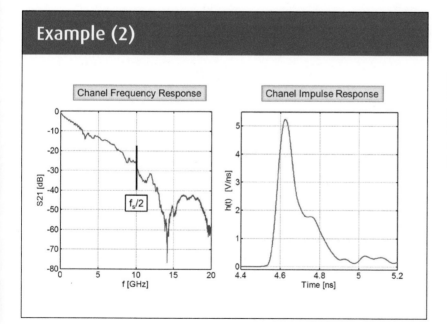

Here is an example of a typical lossy channel along with its impulse response (found via inverse Fourier transform).

25

Shown here are the corresponding convolved impulse response and its second derivative (curvature), to be evaluated at 0.

Example (3)

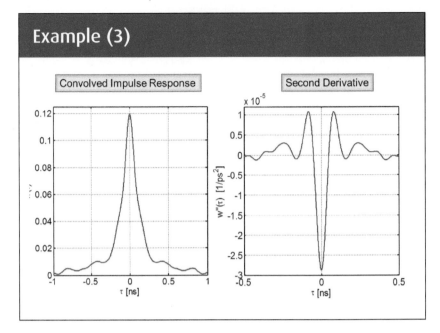

26

For the given channel, the jitter SNR estimate is now 36.4 dB instead of 24 dB (sinusoidal estimate). This makes a very significant difference in terms of clocking requirements (although it may still be "non-trivial").

Example (4)

$$\alpha_t = -\frac{w''(0)}{w(0)} = \frac{2.8 \cdot 10^{-5}}{0.12}\frac{1}{ps^2} = 2.3 \cdot 10^{-4}\frac{1}{ps^2}$$

For $1ps_{rms}$ jitter: $\alpha_t \cdot \sigma_t^2 = -36.4dB$ (OK!)

- Bottom line: the fact that the signal is wideband and filtered by the channel helps, but the jitter spec will still be "non-trivial"

FFE Noise Boosting

27

- Since the noise goes through the FFE along with signal, it will experience noise power boost and PDF expansion.

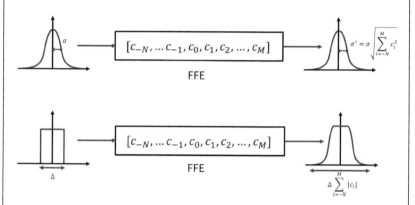

the BER is the final error PDF. Here, we illustrate how different noise PDFs will change after the FFE. For Gaussian noise, the FFE produces another Gaussian PDF with modified noise power. A uniform PDF passed through an FIR filter produces a PDF that asymptotically approaches a Gaussian PDF (Central Limit Theorem). In other words, the total span of the PDF increases as a function of the FFE coefficients. In general, it's the convolution of all scaled uniform PDFs.

Lastly, we consider the impact of feedforward equalization (FFE). FFE boosts noise similar to how it boosts the signal strength. What matters for

Position of FFE – Before ADC

28

- FFE in front of ADC effectively shapes channel first

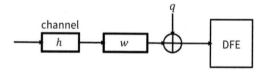

- LSB Δ of ADC at the slicer is determined by
$$\Delta = \frac{2\sum_{i=-\infty}^{+\infty}|h'|_i}{2^B}, \text{ where } h' = h * w$$
- The equalized eye opening is determined by the main cursor strength
$$A = \frac{2h'_0}{PAM - 1}$$
- Then the final signal to quantization error ratio is
$$\frac{A}{\Delta} = \frac{2^B}{PAM - 1} \cdot \frac{h'_0}{\sum_{i=-\infty}^{+\infty}|h'|_i}$$

When FFE is in front of the ADC, we find the signal to quantization error ratio to be proportional to the ADC's nominal resolution, the PAM modulation used, and a channel-dependent term. We will compare this to the digital FFE case next to reach some conclusions.

29

It is more complicated to find a closed form equation for the signal to quantization error ratio in the digital FFE case. As an approximation, we use a noise boost factor α and a pre-cursor degradation on main cursor to model the impairment FFE creates on ADC. The final result also assumes a similar form as the analog FFE case.

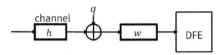

Position of FFE – After ADC

FFE after ADC doesn't shape the channel, and also boosts quantization noise directly

LSB Δ of ADC at the slicer is the following. α is an equivalent noise boost factor

$$\Delta \approx \alpha \cdot \frac{2\sum_{i=-\infty}^{+\infty}|h|_i}{2^B}$$

The equalized eye opening is set w/ a precursor cancellation penalty

$$A \approx \frac{2\left(h_0 - \frac{h_{-1}h_1}{h_0}\right)}{PAM - 1}$$

Then, the final signal to quantization error ratio is

$$\frac{A}{\Delta} = \frac{2^B}{PAM - 1} \cdot \frac{\left(h_0 - \frac{h_{-1}h_1}{h_0}\right)}{\alpha\sum_{i=-\infty}^{+\infty}|h|_i}$$

30

We derive the ADC resolution requirements here. The degradation due to the channel is proportional to the peak to main ratio (PMR) of channel in front. This is different from the peak to average ratio since in link scenarios what matters is channel's main cursor. Placing the FFE before the ADC helps relax the ADC requirement by re-shaping the channel to have a lower PMR. For a perfect channel, PMR = 1 which means there is no channel impairment. For channel with heavy ISI, an

Position of FFE – Comparison

FFE Before ADC	FFE After ADC
$\dfrac{A}{\Delta} = \dfrac{2^B}{PAM - 1} \cdot \dfrac{h'_0}{\sum_{i=-\infty}^{+\infty}|h'|_i}$	$\dfrac{A}{\Delta} = \dfrac{2^B}{PAM - 1} \cdot \dfrac{\left(h_0 - \frac{h_{-1}h_1}{h_0}\right)}{\alpha\sum_{i=-\infty}^{+\infty}|h|_i}$

- Rewriting the terms gives an equation of the following form for both case

$$\log_2\left(\frac{A}{\Delta}\right) = B - \log_2(PAM - 1) - \log_2\left(\frac{Peak}{Main}\right)$$

$$B = B_{Eye} + \log_2(PAM - 1) + \log_2\left(\frac{Peak}{Main}\right)$$

Peak signal to main cursor ratio

$$PMR = \frac{h_0}{\sum_{i=-\infty}^{+\infty}|h|_i}$$

analog FFE can potentially reduce the PMR by more than 2x, which leads to a full 1 bit saving for the ADC.

Position of FFE – SQNR vs. N FFE taps

31

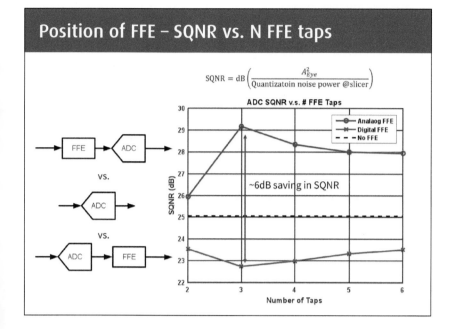

$$SQNR = dB \left(\frac{A_{Eye}^2}{\text{Quantizatoin noise power @slicer}} \right)$$

post-cursor taps. The reference SQNR uses the ADC quantization noise power $\Delta^2/12$ directly. The full scale of ADC is set at the peak value of channel. Digital FFE yields even lower SQNR due to noise boosting. Analog FFE improves SQNR performance significantly due to much lower PMR of equivalent channel. ~6dB SQNR saving is achieved in this ex-

In a MATLAB example, FFE with varying number of taps are used to see how the SQNR changes. One pre-cursor tap is always present, and the rest are ample, and the results will vary depending on the channel. The ADC resolution relaxation saturates with more taps.

Why is This a Big Deal?

32

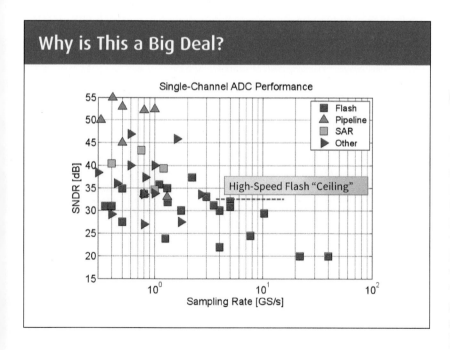

One full bit of savings in the ADC can be very significant in high-speed links. It may mean that a flash topology becomes feasible. Flash ADCs are typically limited to 6 bits or less in resolution but shine in terms of raw (non-interleaved) speed – exactly what we are craving for the most in high-speed applications. The same argument was made in [Rylov, ISSCC 2016].

33

It is possible to get very high speeds with other ADC types through time interleaving (indicated by bold outlines). However, the interleaving factor and thus the complexity will be higher.

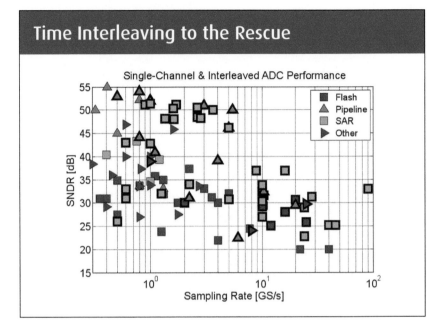

Time Interleaving to the Rescue

34

Shown here is an example that uses 32 interleaved SAR ADCs to get to 28 GS/s. The cost is a complex interleaving network and significant complexity in the signal buffering. The same aggregate speed may be achievable by interleaving only 5-6 flash ADCs. More on this topic will follow later in the ADC section.

State-of-the-Art Example

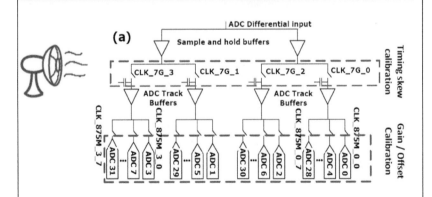

- Massive time interleaving comes with power overhead due to buffering, clock distribution, calibration, etc.

Outline

 35

- Specifying the receiver front-end
  - Resolution requirements
  - Impact of nonidealities
  - Benefits of adding an analog FFE

- Rx FFE design
  - Prior art
  - Inverter-based design
  - Experimental results

- Rx ADC design
  - Performance trends
  - Time interleaving
  - Flash-based design
  - SAR-based design

Objective

 36

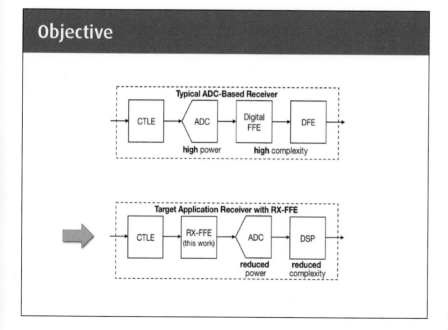

Having shown that an analog FFE may be beneficial, we now show how such an FFE may be implemented. The goal is to generate net savings, i.e. the FFE power should not be higher than whatever we are able to save in the ADC due to the reduced resolution requirement. A potential side benefit is that the digital equalizer power may also be slightly reduced.

37

A common strategy for previous RX-FFE designs was to use inductors to emulate an ideal delay line at a cost of area, as in the example shown here. In addition to the large area, the power consumption for this design is too high to be practical for most transceivers, and the noise is not reported.

Prior Art Example (1/3)

An area of 0.75 mm² due to area-intensive inductors and 80 mW power consumption.

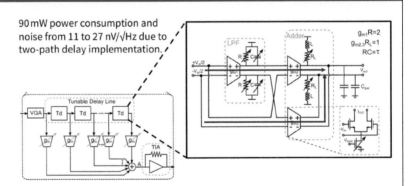

A. Momtaz and M. Green, "An 80 mW 40 Gb/s 7-tap T/2-spaced feed-forward equalizer in 65 nm CMOS," *IEEE J. Solid-State Circuits*, vol. 45, no. 3, pp. 629–639, Mar. 2010.

38

A more recent design implemented a first-order pole/zero Padé approximant using a two-path approach. This method, while an improvement on previous designs, suffers from increased noise and power from the two signal paths in the delay implementation.

Prior Art Example (2/3)

90 mW power consumption and noise from 11 to 27 nV/√Hz due to two-path delay implementation.

E. Mammei, F. Loi, F. Radice, A. Dati, M. Bruccoleri, M. Bassi, and A. Mazzanti, "A power-scalable 7-tap FIR equalizer with tunable active delay line for 10-to-25Gb/s multi-mode fiber EDC in 28nm LP-CMOS," in *ISSCC Dig. Tech. Papers*, pp. 142–143, Feb. 2014.

Prior Art Example (3/3)

 39

Discrete-time FFE complicates the design with power-hungry clocking and ADC-specific implementation.

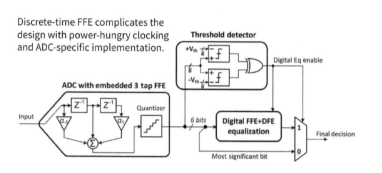

A. Shafik, E. Zhian Tabasy, S. Cai, K. Lee, S. Hoyos and S. Palermo, "A 10 Gb/s Hybrid ADC-Based Receiver With Embedded Analog and Per-Symbol Dynamically Enabled Digital Equalization," in *IEEE Journal of Solid-State Circuits*, vol. 51, no. 3, pp. 671-685, Mar. 2016.

The shown ADC with embedded discrete-time 3 tap FFE is an interesting approach to an FFE implementation. However, it is limited in that it is specific to the ADC architecture and requires a complex and power-hungry clocking network.

Inverter-Based FFE Half-Circuit Schematic

 40

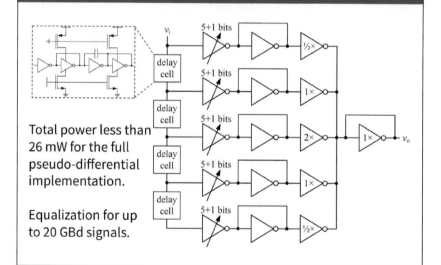

Total power less than 26 mW for the full pseudo-differential implementation.

Equalization for up to 20 GBd signals.

ductors throughout the design enables a high-speed design without the usage of peaking inductors, resulting in a low-area design. The single-path delay (to be introduced later) improves on the two-path delay yielding power and noise improvements. The continuous-time implementation avoids power-hungry and complex clocking networks. As a preview, the proof-of-concept design achieves a 20 GBd symbol rate with less than 26 mW power consumption and less than 10 nV/√Hz spot noise.

The inverter-based FFE presented in this work improves on these previous designs in multiple ways. The usage of high-speed analog-inverter transcon-

41

Analog-Inverter Transconductor

The analog-inverter transconductor is the fundamental building block for the inverter-based FFE. The circuit architecture is identical to that of a digital inverter, but the operating point is constrained to the saturation region in the small-signal range about mid-supply. In this region, the block behaves as a linear transconductor. Due to the class AB operation,

- Biased at half the supply voltage
- Power and noise efficient class AB operation
- Wide linear input range
- High speed without inductors

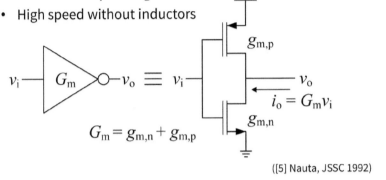

$$G_m = g_{m,n} + g_{m,p}$$

$$i_o = G_m v_i$$

([5] Nauta, JSSC 1992)

this transconductor is efficient in terms of noise, power, and bandwidth while maintaining reasonable linearity performance. This transconductor is a versatile building block and can be used to realize many functions.

42

Analog-Inverter Transconductor Circuits

A simple application of the analog-inverter transconductor is the unity-gain stage, where the self-biased load transconductor behaves as a small-signal resistive load. This stage can easily be modified to realize the FFE summing circuit by adding additional input transconductors and the FFE coefficients by the use of a tunable input transconductor. The addition of a feedthrough capacitor to the unity-gain stage results in a first-order pole and zero Padé delay which is also known as the all-pass delay.

Unity Gain

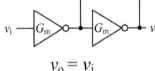

$$v_o = v_i$$

Coefficient

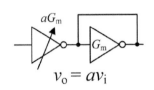

$$v_o = a v_i$$

Summing

$$v_o = v_{i1} + v_{i2}$$

All Pass (Padé Delay)

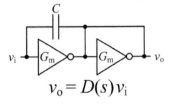

$$v_o = D(s) v_i$$

Padé Delay Concept

43

The Laplace transform of an ideal delay is

$$e^{-s\tau} = 1 - s\tau + \frac{1}{2}s^2\tau^2 - \boxed{\frac{1}{6}}s^3\tau^3 + \cdots.$$

Padé approximants give the best rational approximation of the Taylor series for a given order.

For a first-order pole/zero, the best approximation is

$$\frac{1 - \frac{1}{2}s\tau}{1 + \frac{1}{2}s\tau} = 1 - s\tau + \frac{1}{2}s^2\tau^2 - \boxed{\frac{1}{4}}s^3\tau^3 + \cdots.$$

This is also known as the all-pass delay.

One method to approximate a time delay with poles and zeros is to use the Padé approximant. The Padé approximant gives the best rational function approximation of a desired function in terms of matching the highest possible number of Taylor series coefficients. The left-half plane poles exactly match the right-half plane zeros, canceling in magnitude and summing in phase. Therefore, the magnitude response, in the absence of any additional parasitic poles, is ideal across all frequencies. The phase response is also in good agreement with that of the ideal delay. Because the first-order pole and zero has simple circuit realizations, it has been a popular choice as an analog delay in many previous designs.

Padé Delay Simple Gm-C Realization

44

Pros
✓ Simple Gm-C pole/zero realization
✓ Power efficient and low noise
✓ Zero parasitic nodes

Cons
✗ No high-Z input

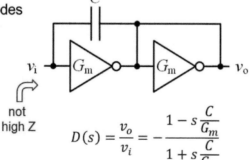

v_i

not high Z

$$D(s) = \frac{v_o}{v_i} = -\frac{1 - s\dfrac{C}{G_m}}{1 + s\dfrac{C}{G_m}}$$

A simple method to achieve the first-order pole and zero in the Padé delay is the Gm-C circuit shown here. It combines the power, noise, and bandwidth performance of the inverter transconductor with the simplicity and equalization performance of the first-order Padé delay. The primary limitation is revealed by analyzing a cascade of two delays, where it is found that the poorly behaved input impedance destroys the prior stage delay characteristics.

45

O ne solution is to in- sert a buffer before each delay to isolate the input impedance, but this adds additional power and the buffer capacitance loads the previous delay output. Therefore, the impact of the buffer stage needs to be carefully consid- ered.

Buffered-Input Padé Delay (1/2)

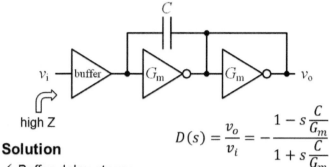

Solution

✓ Buffer delay stages

✗ Costs power and loads previous delay

Requires careful consideration of the buffer realization

$$D(s) = \frac{v_o}{v_i} = -\frac{1 - s\dfrac{C}{G_m}}{1 + s\dfrac{C}{G_m}}$$

46

F or a scaled replica of the gain stage in the delay as a buffer with a scale factor β, the pole and zero are offset by a factor $1+2\beta^{-1}$. For the case of a buffer with in- finite drive (i.e., β→∞), the ideal Padé transfer function is recovered. For the more practical case of equal power in the buffer and the de- lay (i.e. β=1), the pole and zero offset is 3x. Because the power con- sumption of the complete delay is highly dependent on the power consumed in this buffer, it is critical to

Buffered-Input Padé Delay (2/2)

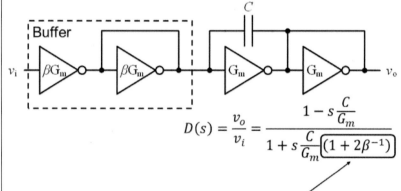

$$D(s) = \frac{v_o}{v_i} = \frac{1 - s\dfrac{C}{G_m}}{1 + s\dfrac{C}{G_m}\boxed{(1 + 2\beta^{-1})}}$$

• Finite output drive results in pole/zero offset
  – How much offset can be tolerated?

understand how much offset between the pole and zero can be tolerated.

Padé Delay Pole/Zero Offset

Consider delays of the form

$$D_\alpha(s) = \frac{1 - \alpha s\tau}{1 + s\tau} \text{ and } D_1(s) = \frac{1 - s\tau}{1 + s\tau}$$

and the corresponding FFE transfer functions

$$H_1(s) = a_0 + a_1 D_1(s) + a_2 D_1^2(s) + \cdots$$
$$H_\alpha(s) = b_0 + b_1 D_\alpha(s) + b_2 D_\alpha^2(s) + \cdots$$

then there exists a linear relation between a and b for each α such that

$$b = M_\alpha a$$
$$H_\alpha(s) = H_1(s).$$

This theorem, presented here without proof, shows that the obtainable equalization transfer functions of an FFE with Padé delays can be exactly replicated by an FFE consisting of any first-order delays. An inherent assumption is that arbitrary coefficient values can be implemented, which is not the case in a practical implementation.

Padé Delay Pole/Zero Offset Example

Example Parameters:

$$\alpha = \frac{1}{3}$$

$$a = \begin{bmatrix} 3 & 2 & 1 \end{bmatrix}^T$$

$$b = M_{\frac{1}{3}} a = \begin{bmatrix} \frac{9}{4} & \frac{3}{2} & \frac{9}{4} \end{bmatrix}^T$$

$$H_{\frac{1}{3}}(s) = H_1(s)$$

- Identical FFE magnitude and phase response
  - For any coefficients, pole/zero offset, and FFE order

Here we present a third-order example to clarify the theorem from the previous slide. As expected, the magnitude and phase response of the FFE with pole and zero offset delays is transformed to exactly match those of the FFE with Padé delays. Although we are demonstrating a specific example, this result can be achieved for any coefficients, pole/zero offset, and FFE order. This result is a consequence the agility of the FFE architecture to equalize in the presence of non-idealities.

49

Coefficient Spread Penalty Bound

The coefficient spread is an important metric because it sets the requirement on the bit resolution of the coefficient's circuit realization. Therefore, it is necessary to understand the effect of the transformation M_α on the coefficient spread. Unfortunately, this transformation is a complicated process with respect to the coefficient spread and no closed-form expression exists. Instead, what we can easily compute is the spectral norm, $\|M_\alpha\|_2$, which gives us the bound plotted here. Note that this bound is not equal to the co-

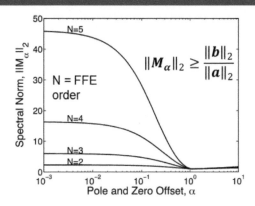

- **Pole and zero offset increases coefficient spread**
  - For $N = 5$ and $\alpha = \frac{1}{3}$, the bound is $10 \geq \frac{\|b\|_2}{\|a\|_2}$

efficient spread, but it is closely related. For N=5 and α=1/3, the bound is $10 \geq \|b\|_2 / \|a\|_2$, but the values observed in practice are more than 2× less than this.

50

Single-Path Padé-Inspired Delay

As a consequence of the theorem and the coefficient spread analysis, we concluded that a reasonable design choice is to burn equal power in the buffer as in the delay, which corresponds to a β=1 and a pole and zero offset α=1/3. The result is the single-path Padé-inspired delay of the inverter-based FFE. Because we started with a low-power G_m-C delay and doubled the power through the addition of a buffer, it is prudent to compare to power performance to an alternative delay implementation.

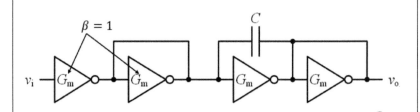

- **Design choice:**
  - Equal power in buffer and delay
  - Pole/zero offset $\alpha = \frac{1}{3}$

$$D(s) = \frac{v_o}{v_i} = \frac{1 - s\dfrac{C}{G_m}}{1 + 3s\dfrac{C}{G_m}}$$

$$\alpha = \frac{1}{3}$$

Two-Path Padé Delay Comparison

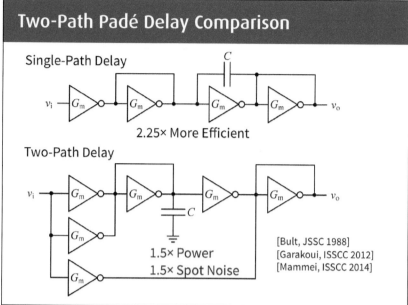

The two-path implementation of the Padé delay is an alternative to the delay introduced in this work. The two-path delay achieves the Padé response by subtracting the input from a first-order pole with a gain of two. A consequence of this approach is that signal is created and then destroyed, wasting power while introducing noise and nonlinearity. The noise is 1.5x the single-path delay noise while using 1.5x power and, therefore, this single-path delay is 2.25x more efficient in terms of noise for the same power. In addition, for the same power (i.e., each G_m in the two-path scaled down by 1.5x), the single-path delay has a stronger output drive by a factor of 1.5x.

Delay Half-Circuit Schematic

- Triode degeneration to compensate for output conductance
- FFE noise determines the delay power consumption
  - Single-ended power of 1.5 mW to limit the FFE output noise to < 1 mV$_{RMS}$

To complete the delay design, triode transistors are used to reduce the load conductance and compensate for the non-zero output conductance of the transconductors. The gates of the triode devices can simply be tied to supply and ground with the residual gain error being absorbed into the coefficients. In a more robust solution, it would be possible to tune the triode gate voltages to adjust for gain and common mode PVT variations. For this design, the device PMOS/NMOS width ratio is chosen to set the common mode to half of the supply voltage. The remaining degree of freedom in the device widths determines the power dissipation and the output noise.

53

Using inverter trans-conductors in the co-efficients and summing circuit results in a ratiom-etric FFE design where the common mode at each stage is common to the entire FFE. In addition, the benefits of the inverter transconductor and gain stage are imparted on the total FFE. The pseudo-dif-ferential implementation is exploited to realize negative coefficients by feeding the signal from the negative half-circuit to the positive coefficient. The input transconductors are a set of binary-weighted min-

Coefficient Half-Circuit Schematic

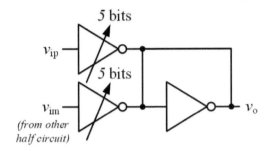

- Two 5-bit binary-weighted analog inverters
  - The second branch enables negative coefficients
  - Coefficient input capacitance loads the delay output:
    $2 \times (2^5 - 1) = 62$ unit inverters

imum-sized inverters that can be switched in or out to achieve the desired coefficient gain.

54

To implement the 5 bits plus sign resolution, $2 \times (2^5 - 1) = 62$ unit inverters are required. This results in a large capaci-tance that loads the de-lay output and limits the bandwidth. To minimize this capacitance, the least-significant bit (LSB) is implemented with se-ries-stacked transistors to effectively lengthen the device by a factor of two. With this modi-fication, the new input capacitance is equivalent to 34 unit inverters which is a reduction of approximately 2x. This comes with a

Coefficient Parasitic Capacitance Reduction

- The LSB is implemented as a $\frac{1}{2} \times$ unit inverter
  - Input capacitance reduced to: $2 \times (2^4 + 1) = 34$ unit inverters
  - ~2 × capacitance reduction
  - Matching penalty for LSB

- Single-ended power between 250 μW and 670 μW

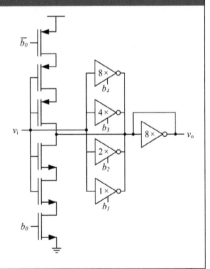

matching penalty, but since it is limited to the LSB, it does not substantially impact the performance.

Summing Circuit Half-Circuit Schematic

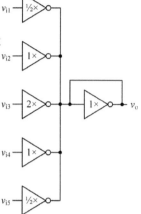

- Scaling factors maximize coefficient dynamic range

- The stage isolates changes in coefficient output conductance

- Single-ended power of 2.9 mW

55

The use of the inverter transconductors in the summing circuit completes the power-efficient ratiometric FFE design. The transconductors are scaled to anticipate the relative magnitudes of the coefficients to maximize the dynamic range of the coefficient resolution.

Die Photo

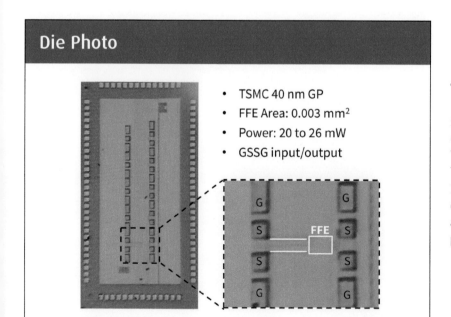

- TSMC 40 nm GP
- FFE Area: 0.003 mm^2
- Power: 20 to 26 mW
- GSSG input/output

56

The performance of this circuit is demonstrated through a proof-of-concept design in TSMC 40 nm GP CMOS. The power is less than 26 mW for all coefficient configurations with an area of just 0.003 mm^2 per FFE.

This slide shows the test PCB used to characterize the proof-of-concept design. The high-speed inputs and outputs are probed through single-ended GSG pads and differential GSSG pads. The low frequency signals (i.e. voltage supply, voltage references, and digital I/O) are bonded directly to the test PCB via the chip-on-board method. In addition, the PCB has a low-speed GPIO interface for scan chain read and write to set the FFE coefficient values and on-chip signal generator amplitude.

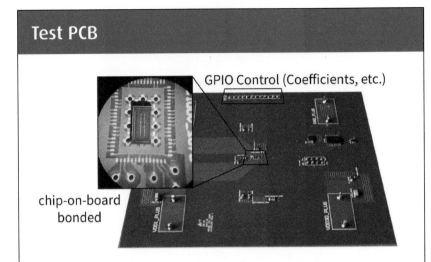

Test PCB

GPIO Control (Coefficients, etc.)

chip-on-board bonded

- Low-frequency signals bonded directly to PCB
- High-frequency signals probed on chip

To characterize the FFE equalization performance, a 50 ps pulse is passed through a channel consisting of a differential 0.5 m FR4 PCB trace. The channel output is probed on the chip and passes through a short 50 Ω terminated on-chip differential transmission line to the FFE input. The pseudo-differential FFE output is driven off chip through GSSG pads and dc blocking capacitors to the high-speed oscilloscope. To adapt the coefficients, the FFE is configured for a single coefficient path by setting all other coefficients to zero. This pulse response is measured on the oscilloscope and the process is repeated for all five coefficient paths to gener-

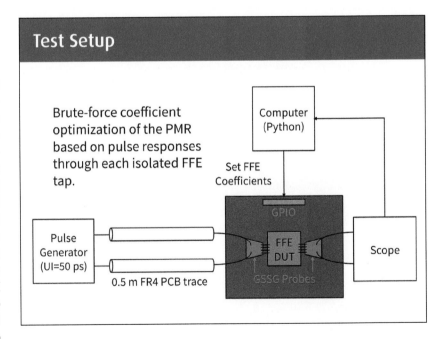

Test Setup

Brute-force coefficient optimization of the PMR based on pulse responses through each isolated FFE tap.

Computer (Python)

Set FFE Coefficients

GPIO

Pulse Generator (UI=50 ps)

FFE DUT

Scope

GSSG Probes

0.5 m FR4 PCB trace

ate a family of pulses. These pulses are processed in MATLAB through the brute-force optimization process to find the optimal coefficient values. These values are then programmed into the FFE on the IC to measure the equalized pulse.

Normalized Pulse Responses

59

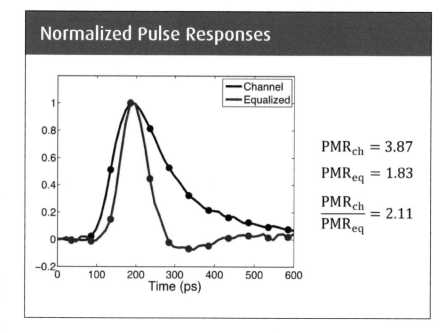

$$PMR_{ch} = 3.87$$

$$PMR_{eq} = 1.83$$

$$\frac{PMR_{ch}}{PMR_{eq}} = 2.11$$

The main cursor attenuation is not depicted and is 3.03× for this channel and set of coefficients. The channel length is limited by the maximum PCB trace length on the test channel board used during testing. Post-layout simulation results suggest that the performance increases for even longer PCB traces with higher channel attenuation.

Normalized PRBS Signals

60

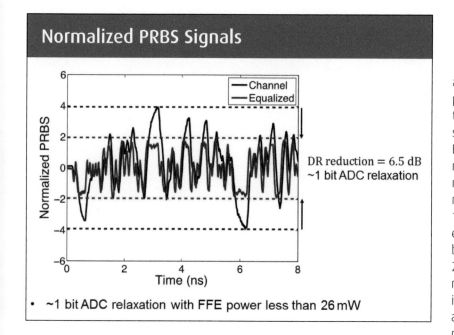

DR reduction = 6.5 dB
~1 bit ADC relaxation

• ~1 bit ADC relaxation with FFE power less than 26 mW

amplitude is equal, the peak amplitude is more than 2× lower for the signal equalized by the FFE. This DR improvement corresponds to a relaxation of the ADC resolution by more than 1 bit. As a point of reference, the 10 GS/s 6 bit ADC in (Zhang, ISSCC 2013) consumes 143 mW of power. This DR improvement results in a reduction of the ADC power by more than 2×

The reduction in signal DR is illustrated in this figure which shows the normalized PRBS responses generated from the associated pulse responses. Although signals are scaled so that the main cursor for a savings of over 70 mW. For this configuration of coefficients, the FFE consumes only 23 mW, resulting in a significant reduction in the total system power.

61

The dominant noise source of the FFE is the delays. Due to the multi-path nature inherent to the FFE architecture, the noise from the delays see multiple paths to the outputs and can add constructively or destructively. The low-noise property of the single-path delay translates into the low measured integrated noise plotted here. The post-layout simulated noise is plotted for comparison and it is in good agreement with the measured data.

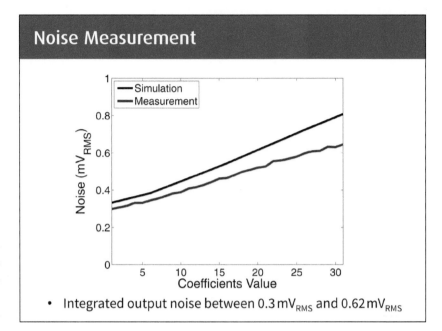

Noise Measurement

- Integrated output noise between $0.3\,\mathrm{mV_{RMS}}$ and $0.62\,\mathrm{mV_{RMS}}$

62

This work achieved a 2× reduction in power per tap while maintaining a competitive symbol rate as compared to previous state-of-the-art designs. Due to the omission of inductors in this design, the FFE area is just 0.003 mm² which is a significant improvement compared to previous designs. The noise is reduced by almost 3× as compared to (Mammei, ISSCC 2014) with no noise numbers reported in (Momtaz, JSSC 2010). The primary sources of these improvements are attributed to the efficient single-path Padé-inspired delay architecture and the efficiency of the analog inverter transconductor.

Performance Summary and Comparison

	This Work	**(Mammei, ISSCC)**	**(Momtaz, JSSC)**
Power (mW)	20 to 26	55 to 90	80
Taps	5	7	7
Power/tap (mW)	**4 to 5.2**	7.9 to 12.8	9.3
Symbol Rate (GBd)	20	10 to 25	40
Process	40nm CMOS	28nm CMOS	65nm CMOS
Supply (V)	1	1	1
Noise (nV/√Hz)	**4.2 to 9.2**	11.4 to 26.6	Not Given
Area (mm²)	**0.003**	0.085	0.75

Outline

 63

- Specifying the receiver front-end
  - › Resolution requirements
  - › Impact of nonidealities
  - › Benefits of adding an analog FFE

- Rx FFE design
  - › Prior art
  - › Inverter-based design
  - › Experimental results

- **Rx ADC design**
  - › **Performance trends**
  - › **Time interleaving**
  - › **Flash-based design**
  - › **SAR-based design**

Speed-Efficiency Tradeoff

64

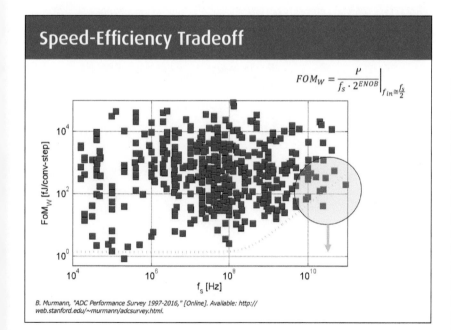

$$FOM_W = \frac{P}{f_s \cdot 2^{ENOB}}\bigg|_{f_{in} \cong \frac{f_s}{2}}$$

B. Murmann, "ADC Performance Survey 1997-2016," [Online]. Available: http://web.stanford.edu/~murmann/adcsurvey.html.

rently working to reduce the energy per conversion-step for ultra-high-speed design and there has been lots of progress in recent years. Like any figure of merit, the Walden FoM is not a perfect measure of efficiency, but it is a good first-order indicator for low-resolution design. See this reference for a discussion of this point: B. Murmann, "The Race for the Extra Decibel: A Brief Review of Current ADC Performance Trajectories," IEEE Solid-State Circuits Magazine, vol. 7, no. 3, pp. 58-66, 2015.

In general, high-speed ADCs are less efficient than low-frequency designs. We use here a plot of the so-called Walden Figure of Merit (FoM) versus speed to show this. High-speed designers are cur-

65

To get a feel for the "velocity" at which the FoM is improving, we consider here the envelope of the data points up until 2011. We see significant progress within a 5-year time frame.

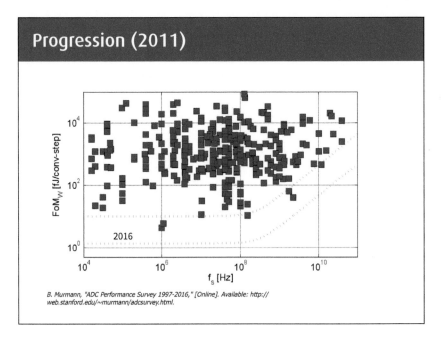

Progression (2011)

B. Murmann, "ADC Performance Survey 1997-2016," [Online]. Available: http://web.stanford.edu/~murmann/adcsurvey.html.

66

Now we limit the data to go back to the year 2006. It is evident that the efficiency of high-speed ADCs has improved by about 100x over 10 years. Converters that used to consume about 10 Watts or more (see e.g. Poulton, ISSCC 2003) are now dissipating only a few hundred milliwatts (see e.g. Kull, ISSCC 2014). Without these significant reductions in power, ADC-based links would be unfeasible. Additional progress is needed to improve the viability of such links further.

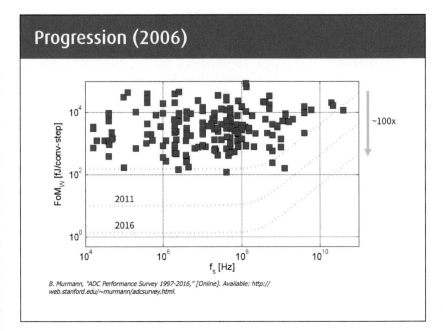

Progression (2006)

B. Murmann, "ADC Performance Survey 1997-2016," [Online]. Available: http://web.stanford.edu/~murmann/adcsurvey.html.

State-of-The Art Designs for High-Speed Links

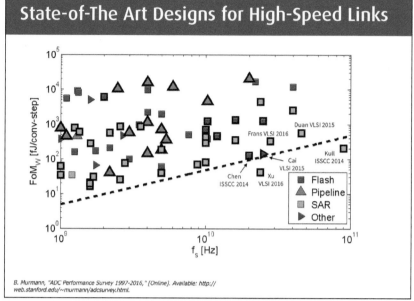

B. Murmann, "ADC Performance Survey 1997-2016," [Online]. Available: http://web.stanford.edu/~murmann/adcsurvey.html.

A zoom into the presented data reveals some of the most efficient high-speed ADCs reported to date. Essentially all relevant designs for high-speed links are time-interleaved, but there exists some di-

versity in the underlying sub-converter architecture. We will look at some of these in more detail later. For the time being, the most promising approach is to time-interleave an array of successive approximation register (SAR) ADCs. For a detailed discussion of this, refer to: B. Murmann, "The successive approximation register ADC: a versatile building block for ultra-low-power to ultra-high-speed applications," IEEE Communications Magazine, vol. 54, no. 4, pp. 78-83, Apr. 2016.

Time Interleaved ADCs: Error Sources

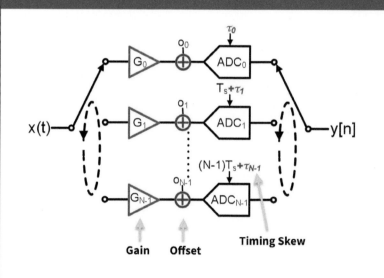

basics. A converter that uses N channels achieves an aggregate sampling rate of N times the sub-ADC conversion rate. The price for this speed increase are various error sources that are introduced by the interleaved structure. The main issues are channel mismatches in gain, offset and timing (sampling instant). There are also errors due to acquisition bandwidth mismatches, but these are often negligible, provided that the nominal bandwidth is significantly large.

Since time interleaving is common to all leading-edge approaches, it is worth reviewing the

69

Compensation of Interleaving Errors

- Gain and offset
  - › Several relatively "simple" techniques exist (usually based on measuring mean and variance in each channel)
  - › Can compensate in analog and/or digital domain

- Timing skew
  - › Much more difficult to handle
    - See overview paper by B. Razavi, JSSC 8/2013
  - › Popular solutions
    - Measure errors in digital domain (many algorithms exist) and compensate via adjustable delay lines → typically incur a jitter penalty
    - All digital measurement & compensation? → Still not practical
    - Measure errors in digital domain, compensate by skewing equalizer taps

70

Gain, Offset and Nonlinearity Cal via Trim-DACs

S hown here is an example of a commercial design and its approach to canceling gain and offset mismatches as well as nonlinearities of the ADC buffers in the analog domain (digital approaches are also in use). At start-up, a calibration DAC drives the input and applies a known signal. The calibration logic computes the gain and offset mismatches and adjusts trim bits to align the channels. Additionally, the ADC reference ladder is skewed to mimic the inverse of the buffer nonlinearity. This latter technique is applicable for flash-based ADCs.

[Verma, ISSCC 2013]

Local Re-Timing & Skew Fine Tuning

 71

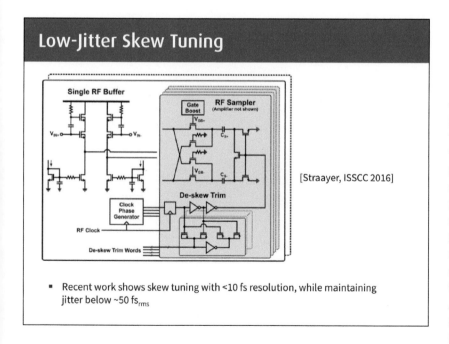

[Kull, VLSI 2013]

it is worth considering local re-timing schemes that minimize the required tuning range. A large tuning range could mean lots of delay cells and a significant jitter penalty. The shown example globally distributes clock c2 to define the sampling instants for all channels. The residual skew is limited by mismatches in the c2 path; the other clocks can tolerate a

In most practical designs, the timing skew is adjusted in the analog domain, usually through tunable delay lines. To limit the complexity of the delay line, large skew. The tolerance for residual skew can be estimated using the jitter equations discussed earlier (see also El-Chammas, TCAS1, 5/2009).

Low-Jitter Skew Tuning

 72

[Straayer, ISSCC 2016]

- Recent work shows skew tuning with <10 fs resolution, while maintaining jitter below ~50 fs$_{rms}$

ing. The shown work achieved 50 fs$_{rms}$ sampling jitter despite skew tuning. This level of jitter was so far typically only seen in non-interleaved designs. At these performance levels, the clock drivers typically consume a large amounts of power and must be carefully optimized. Further reading: A. M. A. Ali et al., "A 14-bit 125 MS/s IF/RF Sampling Pipelined ADC with 100 dB SFDR and 50 fs Jitter," in IEEE Journal of Solid-State Circuits, vol. 41, no. 8, pp. 1846-1855, Aug. 2006.

A major concern among designers is how much jitter an adjustable delay line adds to the clock-

73

Apotentially elegant system solution to the timing skew problem is to absorb it into the back-end equalization. Rather than using one global equalizer, this approach assigns a separate equalizer to each sampling channel. This may not be too unreasonable in practice since: (1) the ADC front-end may only use four switches ("quadrature sampling"), leading to four channels as far as skew is concerned, and (2) the equalizer may anyway need to be split into multiple lanes to facilitate high-speed operation.

Absorbing Skew into the Equalizer

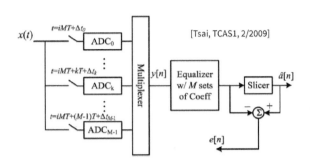

- Split equalizer into M paths and adapt coefficients separately
- For complex equalizers typically found in ADC-based links, one must carefully evaluate if the power overhead is justifiable

74

Power efficient flash-based ADC design must look into making the constituent comparators as small as possible, which means that some form of offset compensation is usually required. Averaging and interpolation techniques are one possible approach, but most recent designs tend to rely on trim-DAC-based offset subtraction. Just like with time skew compensation, a key challenge is to keep the trim DACs small. The previously shown design by Verma

Flash ADC Design

- Resolutions above 6 bits tend to be impractical due to the high comparator count
- Even for 4-6 bits, good power efficiency requires offset calibration (so that small transistors can be used)

et al. (ISSCC 2013) leverages the redundancy of the Wallace encoder to reduce the required trim range.

Offset Calibration at Start-Up

75

- Output oscillates between one and zero as the input-referred offset converges to zero

- Start-up calibration is OK for flash resolutions (temperature drift is manageable)

With a trim DAC available in each comparator, one way to find the proper trim code is to short the circuit's input at start-up and observe the density of zeros and ones at the output. To find the optimum setting, the code is adjusted until the output toggles between 0 and 1 with approximately the same probability. The comparator offset will drift during normal operation (e.g., due to temperature changes), but this effect is often negligible in the low-resolution designs used in high-speed links.

State-of-the-Art Flash-Based Design

76

[Chen, ISSCC 2014]

an aggregate conversion rate of 20 GS/s with good power efficiency. This 6-bit design achieves an SNDR of 30.7 dB at Nyquist frequency and consumes only 69.5 mW (FoMW = 124 fJ/conv-step). Instead of using trim-DACs, this design exploits the randomness of process mismatch to compensate for the clock skew and comparator offset. See the paper for details on the employed statistical element selection approach.

This slide shows the design by Chen et al. (ISSCC 2014). Only 8 flash slices are needed to achieve

77

High-Speed SAR-Slice (8 bits, ~1 GS/s, 22µm x 70µm)

SAR-based designs have evolved as a very strong contender for high-speed design. The reason is that technology scaling has made moderate-resolution SAR ADCs very small and fast. Shown here is an example of a 1-GS/s, 8-bit slice that fits within only 22µm x 70µm. These converter slices contain mostly "digital friendly" components, such as switches, capacitors and latches and will therefore continue to benefit from technology scaling. The speed is upper-bounded by the roundtrip time thought the comparator logic and capacitive DAC. At 1 GS/s, 8 decisions per sample and some extra time for signal acquisition, the effective speed around the loop is about 10 GHz and cannot be improved much further. Scaling will primarily help with power dissipation and area. Maintaining a low metastability rate is also another significant challenge that holds back further slice-speed increases.

[Kull, ISSCC 2013]

Specifications	[1]	[2]	[3]	[4]	[5]	This work		
Architecture	SAR	TI-SAR	TI-SAR	SAR	SAR	SAR		
CMOS Technology (nm)	65	65	65	28	40	32		
Resolution (bits)	8	6	8	8	6	8		
Supply Voltage (V)	1.2	1.2	1.0	1.0	1.0	1.0	1.1	0.9
SNDR near Nyquist (dB)	44.5	31.5	42.75	43.3	30.5	39.3	39.3	38.8
Sampling Speed (GHz)	0.4	1	1	0.75	1.25	1.2	1.3	1.0
Speed per Channel (GHz)	0.4	0.5	0.5	0.75	1.25	1.2	1.3	1.0
Power (mW)	4.0	6.7	3.8	4.5	6.08	3.1	4.2	2.0
FOM (fJ/conf step)	73	210	24	41	178	34	43	28
Area (mm²)	0.024	0.11	0.013	0.004	0.013	0.0015		
Area for 64GS/s (mm²)	3.8	7.0	8.3	0.26	0.67	0.080	0.074	0.096

78

Hierarchical Interleaving (1)

Since many slices are needed to aggregate high-speed with SAR ADCs, the interleaving is often done hierarchically. The show approach is an example where 64 SAR ADCs run in parallel, using a hierarchical 4-4-4 interleaving. Here, the first-rank switches are clocked in quadrature (90 degrees of phase shift between each clock) and the second rank performs a multiplexing operation than fans out the signal by another factor of four. At the last layer, the signal terminates in banks of four SAR ADCs. A disadvantage of this approach is that the high-speed signal must pass through two switches in series. In the design by Kull, this was feasible due to the small parasitics offered by the employed 32nm SOI technology.

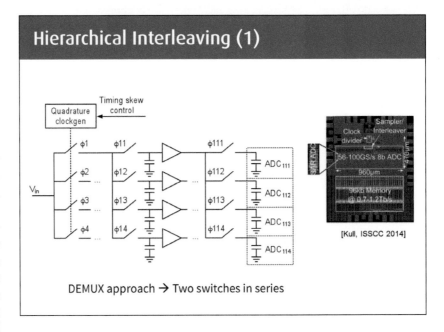

[Kull, ISSCC 2014]

DEMUX approach → Two switches in series

Hierarchical Interleaving (2)

79

(a) ADC Differential input

Sample and hold buffers

CLK_7G_3   CLK_7G_1   CLK_7G_2   CLK_7G_0

ADC Track Buffers   ADC Track Buffers

Timing skew calibration

Gain / Offset Calibration

CLK_875M_3_7 ... CLK_875M_3_0 ... CLK_875M_0_7 ... CLK_875M_0_0

ADC 31, ADC 7, ADC 3, ADC 0, ADC 29, ADC 5, ADC 1, ADC 30, ADC 6, ADC 2, ADC 28, ADC 4, ADC 0

Re-sampling approach → Additional kT/C contribution

An alternative approach is to sample after each buffer. This comes with a penalty of extra kT/C noise; however, this can be managed at the relatively low resolutions demanded by high-speed link applications. Determining the "optimum" allocation of buffers and overall interleaving structure is still an open research problem. The problem can be analyzed similar to logic gate fan-out, except that in this case noise and distortion must also be considered.

Buffer Power Bounds

80

$V_{DD} = 1V$ $g_m/I_D = 10S/A$

g_m
g_m/I_D

I C

$BW = \dfrac{1}{2\pi} \dfrac{g_m}{C}$

$P = V_{DD} \cdot I_D$

Power [W]: 10^{-2}, 10^{-3}, 10^{-4}

$C = 1pF$

Typical regime

$C = 100fF$

$C = 10fF$

BW [Hz]: 10^{9}, 10^{10}, 10^{11}

proportionally to its load capacitance, which is comprised of wiring, sampling capacitances (typically set by noise requirements), junction capacitances etc. In typical designs, the buffer tends to see more than 100 fF of capacitance, leading to power dissipations on the order of 10 mW per buffer, assuming that the MOSFET operates in strong inversion (gm/ID = 10 S/A)

The most common buffer implementation uses a source follower. The power of such a buffer scales for high speed. Minimizing wiring capacitance is a very significant part of minimizing the buffer power.

81

Layout Considerations

Shown here is a good example on an optimized layout of an ADC with hierarchical interleaving. The input node is terminated with 50 Ohms and operates like a transmission line. Its wire length is not critical. The wires after the first buffer are made as short as possible to minimize buffer loading. At this point, there is no 50 Ohm termination and the signal is still not sampled (full bandwidth is seen). The node after the first sampler is less critical, since the signal is now sampled and processed at the reduced sub-ADC rate.

[Straayer, ISSCC 2016]

82

Architecture Variations (1)

Shown on this slide is a design that deviates from the usual flash or SAR approach. This 25 GS/s 8-way time-interleaved uses binary search ADC sub-ADCs and employs a soft-decision selection algorithm to reduce with metastability errors. In 65nm CMOS, the ADC achieves 4.62 bits ENOB at Nyquist and FoMW =143 fJ/conv.-step, while consuming 88 mW and occupying 0.24 mm2 core ADC area.

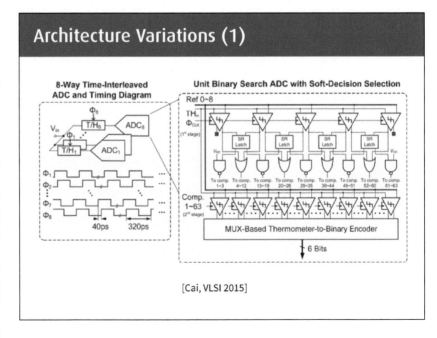

[Cai, VLSI 2015]

Architecture Variations (2)

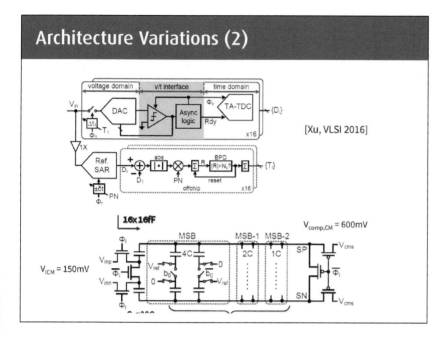

[Xu, VLSI 2016]

The work shown on this slide is a 24GS/s, 6-bit, 16-way time-interleaved ADC featuring a voltage-time (v/t) hybrid two-step structure in its sub-

ADC. The design using a non-hierarchical sampling frontend and hence contains no buffers. Each slice presents only 16 fF of capacitance due to aggressive unit-capacitance scaling. In 28nm CMOS, the ADC consumes 23mW and measures an SNDR/SFDR of 35/54dB for a low-frequency input and 29/41dB for a Nyquist input, respectively. The core area of the ADC is very small, measuring only 0.03 mm². An auxiliary path is used for timing calibration.

Summary

- Wireline communication systems have become a strong driver for the development of low-power, low-to-moderate resolution AFEs running at 10…100 gigasamples per second

- We have come a long way in leveraging the strengths of CMOS to build ever better AFEs
  › But much more work is needed to get to the power levels that future applications will be happy with
  › Both technology scaling and improved design will play a critical role going forward

- Key design aspects
  › Translating system requirement into architectures/circuits
  › Low power clock and signal distribution in interleaved designs
  › Seamless integration of calibration techniques

About the Editors

Kiran Gunnam

Kiran Gunnam currently works as a Technologist at Storage Architecture Research Group of Western Digital Corporation. He was previously Director of Engineering at Violin Memory and also held R&D positions at Nvidia, Blackberry, LSI Corporation, Marvell Semiconductor, Schlumberger, and Intel. Dr. Gunnam is an expert in IC implementation of communications and signal processing systems. His PhD research contributed several key innovations in advanced error correction systems based on low-density parity- check codes (LDPC) and led to several industry designs. He has done extensive work on algorithms and integrated circuits for many applications (including IEEE 802.11n Wi-Fi, IEEE 802.16e WiMax, IEEE 802.3 10-GB, Holographic read channel, Hard-disk drive read channel, and Flash read channel).

Dr. Gunnam has over 60 patents with several others pending. He was an IEEE Solid State Circuits Society Distinguished Lecturer for 2013 and 2014. He received the MSEE and PhD in Computer Engineering from Texas A&M University.

Mohammad (Vahid) VahidFar

Mohammad (Vahid) VahidFar received his M.S. and Ph.D. degrees in Electrical Engineering from Sharif and Tehran Universities in 2002 and 2007 respectively. He worked on data converters and delta-sigma DAC at Marsemai Co for 2004-2005. From 2005, he was with Università di Pavia, as a visiting student working on CMOS reconfigurable receivers for 1-6GHz rang. In 2007, he got a post-doctoral position with Università di Pavia, in cooperation with STMicroelectronics, Pavia, working on UWB receivers and mmW circuits targeting 60GHz. He jointed SiTune Corporation, San Jose, CA in 2009 working on highly integrated TV tuners for cable, mobile and satellite systems. From 2011 he was with Applied Micro Systems, Sunnyvale CA, working on high speed SerDes and CDR circuits for 100Gbps CMOS DQPSK transceivers. He jointed Qualcomm Corp, San Jose CA, in 2013 working on RFIC circuit design for wireless applications including LTE, BT LE and 802.11x standards. He is currently with Apple Inc working on Analog, Mixed signal and RF circuits and systems. Dr. VahidFar has over 25 patents published with several pending patents. He served as Santa Clara SSCS chair for 2014 and 2015.

About the Authors

Bhupendra K. Ahuja

BK Ahuja comes from Qualcomm where he is leading 1Gbps DSL G.fast AFE design. In the past, he has worked for few other companies including Nvidia, Intersil, Teranetics, NeoMagic and Intel designing numerous high performance analog products.

Over 25+ years, his design experiences span Telecom ICs, DSL line drivers, Digital Camera AFE, SERDES and Precision Analog ICs. He holds Bachelor of Technology (EE) from I.I.T., Kanpur, India , Master's and Ph.D. (EE) from Carleton University, Ottawa, Canada. BK has published over 25 technical papers and holds numerous patents in the area of Analog IC designs. BK is the inventor of "Ahuja Compensation" a novel frequency compensation technique widely used for CMOS operational amplifiers. In 2006, he was elected IEEE Fellow for his contributions to analog IC designs.

Ryan Boesch

Ryan Boesch completed his bachelor's degree at Iowa State University in 2008, Summa Cum Laude, and Master's degree in 2009. He received his PhD degree at Stanford University in 2016. He has interned at multiple IC design companies including Micron Technology, Linear Technology, Texas Instruments, and Oracle Labs. He is presently working at Raytheon Space and Airborne Systems in Santa Barbara, CA.

Thomas H. Lee

Professor Tom Lee leads Stanford Microwave Integrated Circuits Laboratory. Thomas H. Lee received the S.B., S.M. and Sc.D. degrees in electrical engineering, all from the Massachusetts Institute of Technology in 1983, 1985, and 1990, respectively. Professor Lee's principal areas of professional interest include analog circuitry of all types, ranging from low-level DC instrumentation to high-speed RF communications systems. His present research focus is on CMOS RF integrated circuit design, and on extending operation into the terahertz realm.

He served for a decade as an IEEE Distinguished Lecturer of the Solid-State Circuits Society, and has been a Distinguished Lecturer of the IEEE Microwave Society as well. He holds approximately 70 U.S. patents and authored The Design of CMOS Radio-Frequency Integrated Circuits and Planar Microwave Engineering, both with Cambridge University Press. He is a co-author of four additional books on RF circuit design, and also cofounded Matrix Semiconductor (acquired by Sandisk in 2006). He founded ZeroG Wireless (acquired by Microchip) and is a cofounder of Ayla Networks. He served as MTO Director at DARPA from April 2011 to October 2012.

In early April of 2011 he was awarded the Ho-Am Prize in Engineering (colloquially known as the "Korean Nobel"), and in 2012 he was awarded the U.S. Secretary of Defense Medal for Exceptional Civilian Service for his work at DARPA.

Omeed Momeni

O meed Momeni received the B.Sc. degree from Isfahan University of Technology, Isfahan, Iran, the M.S. degree from University of Southern California, Los Angeles, CA, and the Ph.D. degree from Cornell University, Ithaca, NY, all in Electrical Engineering, in 2002, 2006, and 2011, respectively.

He joined the faculty of Electrical and Computer Engineering Department at University of California, Davis in 2011. He was a visiting professor in Electrical Engineering and Computer Science Department at University of California, Irvine from 2011 to 2012. From 2004 to 2006, he was with the National Aeronautics and Space Administration (NASA), Jet Propulsion Laboratory (JPL), to design L-band transceivers for synthetic aperture radars (SAR) and high power amplifiers for Mass Spectrometer applications. His research interests include mm-wave and terahertz integrated circuits and systems.

Prof. Momeni is the recipient of National Science Foundation CAREER award in 2015, the Professor of the Year 2014 by IEEE at UC Davis, the Best Ph.D. Thesis Award from the Cornell ECE Department in 2011, the Outstanding Graduate Award from Association of Professors and Scholars of Iranian Heritage (APSIH) in 2011, the Best Student Paper Award at the IEEE Workshop on Microwave Passive Circuits and Filters in 2010, the Cornell University Jacob's fellowship in 2007 and the NASA-JPL fellowship in 2003.

Boris Murmann

Boris Murmann (S'99–M'03–SM'09–F'15) received the Dipl.-Ing. (FH) degree in communications engineering from Fachhochschule Dieburg, Dieburg, Germany, in 1994, the M.S. degree in electrical engineering from Santa Clara University, Santa Clara, CA, USA, in 1999, and the Ph.D. degree in electrical engineering from the University of California, Berkeley, CA, USA, in 2003.

From 1994 to 1997, he was with Neutron Mikrolektronik GmbH, Hanau, Germany, where he developed low-power and smart-power ASICs in automotive CMOS technology. Since 2004, he has been with the Department of Electrical Engineering, Stanford University, Stanford, CA, USA, where he is currently an Associate Professor. His research interests are in the area of mixed-signal integrated-circuit design, with special emphasis on data converters and sensor interfaces.

In 2008, Dr. Murmann was a co-recipient of the Best Student Paper Award at the VLSI Circuits Symposium and the recipient of the Best Invited Paper Award at the IEEE Custom Integrated Circuits Conference (CICC). He received the Agilent Early Career Professor Award in 2009 and the Friedrich Wil-helm Bessel Research Award in 2012. He served as an Associate Editor of the IEEE Journal of Solid-State Circuits and as the Data Converter Subcommittee Chair of the IEEE International Solid-State Circuits Conference (ISSCC). He is the pro-gram chair for the ISSCC 2017.

Nikola Nedovic

Nikola Nedovic is a senior research scientist at NVIDIA Corp., Santa Clara, CA. He received a Dipl.Ing. degree in electrical engineering from the University of Belgrade, Serbia, in 1998 and the Ph.D. degree from the University of California at Davis, in 2003.

In 2001, he joined Fujitsu Laboratories of America, Inc., Sunnyvale, CA, where he worked on high-speed communications and high-performance and low-power circuits for electrical and optical communications. In 2016 he joined NVIDIA Research where he works on system and circuit design for short reach low power high speed links. His research interests include high-speed analog and mixed-signal circuits and system modeling for wireline communications, clock and data recovery, and circuit design and clocking strategies for high-performance and low-power digital applications.

Kevin Zheng

Kevin Zheng received B.S. degree in electrical engineering and computer science (EECS) from the Massachusetts Institute of Technology (MIT), Cambridge, MA in 2012. He received Master of Engineering degree in EECS from MIT through VI-A industry program with Analog Devices Inc. in 2013. In 2013, Kevin received the J. Francis Reintjes Excellence in VI-A Industrial Practice Award and the David Adler Memorial EE MEng Thesis Award.

Currently, Kevin is a Ph.D. student in the Murmann Mixed Signal group. His research interests include high-speed mixed-signal circuit design and system level analysis methodology, specifically for ADC-based high-speed links.